S0-BWR-256

Women's Renunciation in South Asia

Religion/Culture/Critique

Series Editor: Elizabeth A. Castelli

How Hysterical: Identification and Resistance in the Bible and Film
By Erin Runions
(2003)

Connected Places: Region, Pilgrimage, and Geographical Imagination in India
By Anne Feldhaus
(2003)

Representing Religion in World Cinema: Filmmaking, Mythmaking, Culture Making
Edited by S. Brent Plate
(2003)

Derrida's Bible (Reading a Page of Scripture with a Little Help from Derrida)
Edited by Yvonne Sherwood
(2004)

Feminist New Testament Studies: Global and Future Perspectives
Edited by Kathleen O'Brien Wicker, Althea Spencer Miller, and Musa W. Dube
(2005)

Women's Renunciation in South Asia: Nuns, Yoginis, Saints, and Singers
Edited by Meena Khandelwal, Sondra L. Hausner, and Ann Grodzins Gold
(2006)

Retheorizing Religion in Nepal
By Gregory Price Grieve
(2006)

The Religious Dimensions of Advertising
By Tricia Sheffield
(2006)

Gender, Religion, and Culture in the Pre-Modern World
Edited by Brian Britt and Alexandra Cuffel
(2007)

WOMEN'S RENUNCIATION IN SOUTH ASIA

NUNS, YOGINIS, SAINTS, AND SINGERS

Edited by

MEENA KHANDELWAL,
SONDRA L. HAUSNER, AND
ANN GRODZINS GOLD

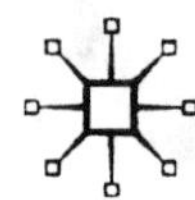

WOMEN'S RENUNCIATION IN SOUTH ASIA

First published in 2006 by
PALGRAVE MACMILLAN™
175 Fifth Avenue, New York, N.Y. 10010 and
Houndmills, Basingstoke, Hampshire, England RG21 6XS
Companies and representatives throughout the world.

PALGRAVE MACMILLAN is the global academic imprint of the Palgrave Macmillan division of St. Martin's Press, LLC and of Palgrave Macmillan Ltd. Macmillan® is a registered trademark in the United States, United Kingdom and other countries. Palgrave is a registered trademark in the European Union and other countries.

ISBN-13: 978–1–4039–7221–7
ISBN-10: 1–4039–7221–4

Library of Congress Cataloging-in-Publication Data

Women's renunciation in South Asia : nuns, yoginis, saints, and singers / edited by Meena Khandelwal, Sondra L. Hausner, and Ann Grodzins Gold.
p. cm.—(Religion/culture/critique)
Includes bibliographical references and index.
Contents: Introduction / Sondra L. Hausner and Meena Khandelwal—Do saints need sleep? / Meena Khandelwal—Living practical dharma / Sara Shneiderman—The true river Ganges / Kristin Hanssen—Staying in place / Sondra L. Hausner—Passionate renouncers / Kalyani Devaki Menon—How Buddhist renunciation produces difference / Kim Gutschow—Renouncing expectations / Lisa I. Knight—These hands are not for henna / Anne Vallely—Afterword : breaking away— / Ann Grodzins Gold.
ISBN 1–4039–7221–4
1. Women—Religious life—South Asia. 2. Women and religion—South Asia. 3. Monastic and religious life of women—South Asia. I. Khandelwal, Meena. II. Hausner, Sondra L. III. Gold, Ann Grodzins, 1946–. IV. Series.

BL1055.W66 2006
206'.570820954—dc22 2005056645

A catalogue record for this book is available from the British Library.

Design by Newgen Imaging Systems (P) Ltd., Chennai, India.

First edition: August 2006

10 9 8 7 6 5 4 3 2 1

Printed in the United States of America.

To Tessa Bartholomeusz (1958–2001) and
Lynn Teskey Denton (1949–1995)
whose pioneering ethnographic research
with women renouncers in Sri Lanka and India
inspires and grounds this work.

Contents

List of Illustrations

Series Editor's Preface

Religion/Culture/Critique is a series devoted to publishing work that addresses religion's centrality to a wide range of settings and debates, both contemporary and historical, and that critically engages the category of "religion" itself. This series is conceived as a place where readers will be invited to explore how "religion"—whether embedded in texts, practices, communities, or ideologies—intersects with social and political interests, institutions, and identities.

Women's Renunciation in South Asia: Nuns, Yoginis, Saints, and Singers offers ethnographically rich and detailed portraits of women living according to ascetical imperatives expressed in Buddhist, Hindu, Muslim, Jain, Himalayan Bon, and Bengali Baul traditions across South Asia. The project focuses on some of the most pressing theoretical concerns of contemporary religious and cultural studies: the importance of studying lived religion and its complex interactions with received textual traditions; questions of globalization, transnationalism, and comparison; figurations and lived experiences of gender, the body, and sexuality; and religious practice itself as a potent cultural *and* counter-cultural mode of expression and critique. This collection of essays adds depth and resonance to—and ultimately creatively disrupts—the contemporary representation of South Asian religious life by drawing readers' attention to a series of carefully drawn portraits of the quotidian lives of individual women renunciants in India, Nepal, and Bangladesh. Generalizations about "South Asian religion" recede, at least long enough for complicating questions, unsettling and messy details, and new frames of reference to enter in. Crossing national borders, transgressing the boundaries that so often separate the study of particular religious traditions, and foregrounding the ethnographic relationships out of which these essays have emerged, *Women's Renunciation in South Asia* contributes meaningfully to the contemporary project of fieldwork-based research in religious studies and adds crucial facets to the conversations that this series seeks to inaugurate and support. Thanks to editors Meena Khandelwal, Sondra L. Hausner, and Ann Grodzins Gold for their labors in bringing these remarkable essays together.

Elizabeth A. Castelli
RELIGION/CULTURE/CRITIQUE Series Editor

New York City
December 2005

Acknowledgments

First and foremost, we thank the women renouncers whose lives are presented in these chapters. They let us into their lives and patiently tolerated our sometimes clumsy attempts to understand their worlds. Second, we recognize the contributors to this volume as true ethnographers of South Asia, able to spend years with families and communities foreign to them, and to communicate those experiences with depth and sensitivity in their writing. Each has a story to tell, based on solid and meaningful relationships built over time. We admire the way each of our contributors lived in and represents the field, and we thank them for their ideas, encouragement, and willingness to respond to our endless requests. We are grateful to our families who supported us even during our absences from other responsibilities as we neared the completion of this book. Finally, we appreciate the institutional support offered by the Departments of Women's Studies and Anthropology at the University of Iowa, including Carrie Freeman and Lavanya Murali's able and crucial research assistance, the Save the Children-U.S. Himalayan Field Office, and the Department of Religion and College of Arts and Sciences of Syracuse University. Mark Hauser skillfully produced our map and must be thanked not only for terrific cartographic expertise but for his patience. Our heartfelt thanks also to Laurie Winship for her meticulous labor on the index.

It is no easy task to edit a book three ways, across two continents. But the wealth of experience our contributors and informants offer has meant that thinking through these essays together—across time zones, under deadlines, over hundreds of emails, and via thousands of air miles—has been an exceptionally rich intellectual experience that none of us could pass up. Our hope is that putting them together in one volume offers students and scholars of religion, gender, and South Asia new food for thought and new angles through which to see these worlds.

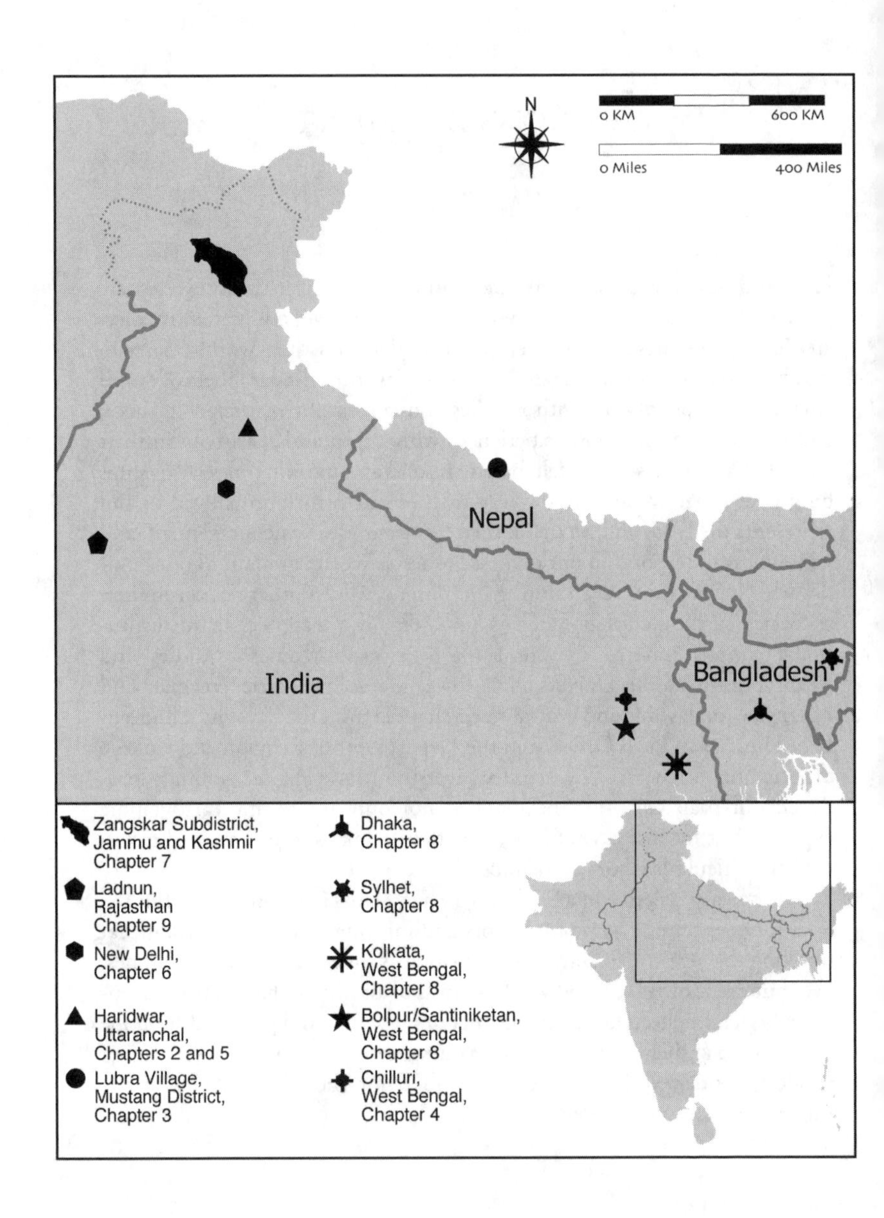
N
0 KM
600 KM
0 Miles
400 Miles
Nepal
India
Bangladesh
Zangskar Subdistrict, Jammu and Kashmir Chapter 7
Ladnun, Rajasthan Chapter 9
New Delhi, Chapter 6
Haridwar, Uttaranchal, Chapters 2 and 5
Lubra Village, Mustang District, Chapter 3
Dhaka, Chapter 8
Sylhet, Chapter 8
Kolkata, West Bengal, Chapter 8
Bolpur/Santiniketan, West Bengal, Chapter 8
Chilluri, West Bengal, Chapter 4

Chapter 1

Introduction: Women on Their Own

Sondra L. Hausner and Meena Khandelwal

The women who are the primary subjects of this book live unusual lives. They have broken away from the social roles and structures expected of them, in each of their various cultures. First and foremost, they are religious practitioners. Their primary social networks and the ways in which they relate to the world around them are charted through the lenses of religious devotion and duty, rather than the template of public expectations about feminine behavior. Their primary goals are liberation, enlightenment, and freedom from the cycles of *samsara*, the wheel of birth and death. In addition to the meditative or yogic disciplines they undertake to attain their religious ideals, they must tend to the exigencies of real life, the daily requirements of food and shelter for themselves and those who depend on them.

Most of the women in this book live on their own, rather than as wives and mothers. They may still live in family settings, but family is constructed differently for renouncers than it is in most popular interpretations of the hallowed concept. Some of these women live in ashrams or monastic communities that feel like families and may even serve as sources of financial support. Others still rely on natal families for material support but live in residential institutions for the training of nuns. In some cultures, women who decide to live alone or recreate a sense of family in a broader context are not unusual. In South Asia, where the women in this book live, the social expectation that women will marry and procreate is overwhelming, and choosing to live as a nun, *yogini*, or *sadhvi*—a female renouncer—is considered foolish, unwise, or downright defiant. Women who choose such a path must somehow be slightly touched, people say, or extremely gifted, or both. Without exception, the lay or householder communities around these renouncers believe them to be powerful women.

This book tells women's stories from their own perspectives, in order to show how real-life renouncers in South Asia negotiate these difficult but important roles. Each woman in this book has made certain choices, or has had choices made for her, that require an independence of spirit and a willingness to face social projections or judgments that are not always easy or pleasant. As a group, women who renounce in South Asia do so because of their respective faiths in the gods and the gurus they follow, and their willingness to bear social approbation, isolation, or a life without luxury. After all, asceticism—in most traditions around the world and for women as well as for men—requires doing without, and accepting what comes. But being a woman means that cultural expectations are all the more confining, and the steps required to break free from traditional roles all the more challenging.

Certain life experiences are common to the renouncers on these pages: they are all women who live separately from so-called traditional South Asian social structures and conventional social roles. At the same time, the places, contexts, and ways in which they live vary widely. Although we do not claim to cover all traditions or locations in South Asia in this volume,[1] we describe women who live in three different countries, and ten different regions within those countries. Some are based in urban settings, some in rural settings, and some move frequently between locations. Some live in institutions, placed there by family members, while others have fled their natal communities and live entirely self-sufficiently. Some perform their dharmic or religious duties in local community settings, while others work for specific agendas at local, national, or regional levels.

In addition, the women renouncers whose lives we recount follow multiple religious paths. South Asia as a historically constituted region has given birth to and nurtured a rich variety of ascetic traditions. In this volume, we consider the realities of women who follow Himalayan Buddhism, Himalayan Bon, Jainism, Islam, Bengali Baul practice, and Hinduism. Each of these religious traditions has certain models of practice and specific rituals and ritual behavior that renouncers are trained to conduct. The model of renunciation—what renouncers are supposed to do, how they are supposed to behave, and who they are expected to engage with—looks different in each religion, but there are also certain common threads. Religious traditions are historically linked, and the roles and expectations of renouncers in different monastic orders may derive from common textual sources as well as public, everyday cultural practices. This volume explores the ways in which women following different South Asian religious paths share common experiences, despite the variation in their individual practices and the wide range of ways in which renouncer and lay communities engage with each other.

Being a woman, and particularly being a woman renouncer, may be a stronger common thread of experience than being a Hindu, a Muslim, or a Buddhist. For all their regional and religious diversity, the women in this book share certain conundrums. They all face the question of defining community, or establishing the company they will keep, and their relationship to that community. All have wondered whether to marry, remarry, or stay married, and have struggled with how to negotiate the unquestioned South Asian social value of having a husband and being a wife. They have been trained in appropriate religious disciplines—different from the household rituals lay women perform—and each has had to make decisions about how to balance the requirements of solitary pursuit with those of social service and engagement. They continually face the questions of how they will sustain themselves and those who rely on them. Each has had to find, follow, and adhere to the religious instructions of her guru, not always without complications. And all face the question of what to make of—and how to handle—their women's bodies and the desires, fears, and emotions that embodiment entails.

This book traces common themes in the lives of contemporary women renouncers by looking at the choices they have made and the paths they have chosen to follow. In some cases, they aspire to and even resemble textbook ascetics. The popular imagination in both South Asia and the so-called West tends to characterize renouncers as belonging to monolithic social orders, in part because they have made such radical life choices in relation to their respective communities, and in part because the Christian monastic tradition is so closely tied to institutions that dictate strict behavior. But part of the premise of this volume is that everyday experiences vary widely among women everywhere. When we look closely at the life circumstances and the realities of living in late twentieth century South Asia for each of the women with whom we worked, we find that their biographies and goals are extraordinarily diverse: they come from different parts of South Asia, are trained in various traditions, enjoy different levels of privilege, and engage in multiple practices. Even within traditions, renunciation and religious practice look different to every woman who pursues them.[2]

Renunciation in South Asia: Text and Practice, Men and Women, Monastics and Laity

Renunciation has a long and noble history in South Asia, spanning the major religions of the region. In the Hindu tradition, a good deal of textual material outlines how renouncers should live, what they should avoid, and the religious goals they should aspire to achieve (cf. Olivelle 1992; Doniger and Smith 1991). Similarly, in the Tibetan Buddhist tradition, a series of precepts have been translated from Sanskrit texts that specify appropriate conduct for nuns and monks (cf. Holmes and Holmes 1995). The teachings of the Jain teacher Mahavira led to the development of detailed works on monastic discipline, as well as texts that precisely dictate how aspirants should understand the relationship between body and soul (cf. Tatia 1994).[3] Textual dictates vary among religious traditions, of course, and even within them, from one historical period to another.

Taken as a group, textual doctrines often influence popular perceptions of how ascetics live in South Asia today. The text-based images of a somber nun in robes spending her days in meditation practice, or of an isolated yogi—almost always male—living high in a Himalayan forest, are prevalent both in South Asia and in the so-called West (cf. K. Narayan 1993). But the real lives of renouncers invariably look quite different from textual ideals. And this is especially true in the case of women renouncers, since the texts are generally written by, for, and about men (cf. Denton 2004; Simmer-Brown 2001; Wilson 1996; Olivelle 1992; U. King 1984). Women are almost always the dangerous seducers in these texts, rather than the renouncers themselves.[4] In this volume, however, women are not the taunting demons of lore, who threaten to lure well-intentioned but susceptible yogis to the fires of hellish delusion (cf. O'Flaherty 1980). They are themselves the practitioners who must negotiate the rocky terrains of desire and attachment, family and sexuality, ego and expectation, and the particularly thorny experience of embodiment.

In both text and reality, for both men and women, and in all religious traditions, renunciation is about confronting social reality alongside material reality and spiritual aspiration. For all the women in this book, religious life involves a rejection of social ideals and conventional practices, sometimes on more than one level. As Knight (this volume) writes, the act of renunciation is as much about society as it is about religion. In the context of the Bengali Baul tradition, she tells us that "renunciation for Bauls actually has more to do with society than it does with the Baul path itself." Renunciation is, for many of the women in this book, about mitigating social, economic, and religious circumstances that make life as a solitary woman, or single mother, or poor widow, very difficult. Ironically, the path of renunciation can make life in the social world a little easier.

Contemporary renunciation in South Asia is one way to sidestep social hierarchies that can be experienced as rigid or unyielding, including gender and caste.[5] For this reason, many (although not all) of the women we worked with come from lower Hindu castes or disadvantaged groups.[6]

Hanssen's work with members of a low-caste Baul family in West Bengal, for example, and Gutschow's work with very poor communities in remote Zangskar show us how renouncing the social world can be a practical escape from marginality, or from impossible material circumstances. Renunciation is ideologically about leaving behind social distinctions, so a low caste, severe poverty, an inability to integrate, or an unwillingness to follow social norms are all good reasons for women to renounce the world that created these injustices.

Gender is one of the most persistent and pervasive social divisions, and as for caste, we see in this volume numerous instances of how renunciation can be used to sidestep difficult life circumstances. In Knight's work, for example, we see how single Baul women renounce partly to avoid the stigma attached to their unmarried status, and in Hausner's, we see how women might renounce in order to leave a bad marriage or an unsupportive home setting. For all the women in this book, the act and practices of renunciation are in part about attempting to free oneself from the ideological—and sometimes theological—constructs that undermine the female gender so strongly in Hindu, Buddhist, and Jain traditions. And for most, renunciation is one of the more legitimate ways to either quietly avoid or reject outright the powerful cultural expectations that all women will become docile wives and perfect mothers.

Interacting with Lay Communities

The renouncers in this book have used their religious affiliations and practices to forge social worlds that provide alternatives to the natal families most wished to leave behind, and the marital families most wished to avoid. Although renunciation often requires leaving families and communities, the path is rarely antisocial.[7] The women we describe regularly interact with audiences, disciples, gurus, village residents, and fellow members of monastic communities. In this volume, Menon describes how *sadhvis* sometimes preach to large audiences of devout followers and potential converts to Hindu nationalism, and Khandelwal shows us how they may establish sizeable ashrams that undertake social welfare projects to serve the needs of their lay followers. Shneiderman tells us how an aged renouncer in the remote village of Lubra, northern Nepal, was sometimes asked to conduct rituals in village households; she dreams of creating a monastic school that would educate the young girls of her area in Bon religious practice. The women in this volume recreate communal settings in many ways—through interactions with local residents, patrons, sponsors or donors, or ritual clients. Relationships with the laity can mean either wholehearted support or ignoble rejection for renouncers in householder communities.

Despite ideologies of solitude, family structures—adoptive families, recreated families, and religious orders—may also step up to provide the kind of emotional intimacy and material support families ideally assure. The ashram Khandelwal describes feels like an extended family, where members are together supported and directed by a singular *sadhvi* whose mission is as much social support as it is spiritual guidance. The single woman with whom Hausner worked raises a child that is not her own, as a way of serving the community in which she lives. And although most South Asian religious ideologies and doctrines emphasize explicit departure from family life, both marital partners and biological parents sometimes remain literal sources of support. Marital families are core to the lives of some Bauls, Bengali ritualists who may use partners in practice (Hanssen, this volume), and natal family members must continue to provide financial support for nuns in Zangskar, where monastic institutions for women do not receive enough public funding (Gutschow, this volume).

Communal support also comes through peers in a monastic residential community or ashram: not all women renouncers are closely involved with lay communities. Indeed, Jain nuns deliberately keep apart from the explicit caring for others, as Vallely describes in this volume. They take meticulous care not to harm any living creature in their actions, and they live in communal settings designed for the mutual support of renouncers. But these institutional settings are ideologically, socially, and spatially separated from those of the laity. As Vallely writes, "Because the purpose of the *sadhvi's* life was not tied up with the caring and comforting of others, she was free to sever her ties with the world. Unlike her lay sisters, no earthly responsibilities beckoned her, nor did they take precedence over her own spiritual needs." This explicit and intended split with the laity serves to focus each nun's endeavor on renunciation alone, but the sisterhood of Jain nuns endures as a form of social support.

In none of the essays in this book is renunciation an entirely isolated endeavor. Quite apart from any social interactions renouncers may have with fellow monastics, family members, or other members of the laity, the essays in this book are all based on renouncers' encounters with ethnographers who chose to follow their every move. In the context of religious life where the company one keeps is believed to influence one's energetic and social levels, interactions with society have very specific meanings and come out of specific choices. All of these social encounters—with family or community members, peers, and ethnographers—are the source of much reflection for the women discussed in this book, because they know that the path of renunciation is about leaving society behind. Each has reinterpreted her relationships with the world, and created new ways of interacting with members of natal, marital, and adopted communities.

The Institutions of Renunciation

Ashrams and monastic communities are plentiful in South Asian Hindu, Buddhist, and Jain traditions, and women's institutions are an accepted subset. The existence of an alternative communal setting means that, through the language of religion, women who have nowhere else to turn can strike out on their own. This may mean that women who want to leave their natal or marital homes can join or create an ashram, as did some of the Hindu women with whom we worked (cf. Khandelwal and Hausner, both this volume), or that women can leave a monastic setting for an isolated retreat when the need for greater practice calls, as did some of the Buddhist women in this book (cf. Gutschow, this volume). The religions represented in this volume take departure from one's natal environment seriously, for both spiritual and social reasons, and the acts of leaving, detaching, and moving on are integral to the theological traditions women renouncers follow. Ironically, the ideology of solitude for renouncers in South Asian religions means that women who feel alone have a place to turn, and a somewhat legitimate alternative structure to join.

Religious life offers a common bond to women who leave home, and creates an alternative community, whether that community is a nunnery, an ashram, or a loosely organized network of like-minded renouncers. A core principle of the anthropology and sociology of religion is that common religious practice binds people together and creates a new kind of society (Durkheim1995[1912]). In this volume, we see this dynamic of communal religious life in the work of Vallely, who works within the sisterhood of Jain nuns, mutually supportive in their theology and practice, and of Khandelwal, who describes a bustling ashram environment where being a devotee of the same Hindu guru justifies collective living with virtual strangers, not to mention long hours of hard work in the kitchen to prepare large feasts. Menon captures a somewhat different dynamic in her account of Hindu nationalist *sadhvis*: in analyzing their rhetoric, we see how the ideology of a religious community can be used for explicitly political ends.

Although individual renouncers are sometimes considered outcasts of a dominant or normative social system, most institutions of renunciation are supported by the laity. Individual renouncers are also supported by alms and donations from laypersons who offer money in the spirit of respect, or to obtain merit or a blessing.[8] The Baul renouncers we describe in this book sing devotional songs for a living, traveling up and down train compartments to solicit donations from lay passengers (Hanssen and Knight, both this volume). Religious practitioners are considered those who do the work of faith for an extended community, a popular belief that in turn funds a monastic institution in its different forms. This exchange of material support for ritual action or philosophical discourse is the basis of renouncer-householder relations throughout South Asia.[9] Whether one wealthy patron donates a large sum to a women's ashram, or individual pilgrims offer alms to begging yogis, collective understandings that faith is productive, and that renouncers serve a purpose for the larger social whole, permeate South Asia.[10]

In this model, institutions of renunciation are productive both for the spiritual well-being of a larger community (an irony when we consider that individual renouncers may be simultaneously supported and outcast), and for the individuals who benefit from the accumulated resources of a religious institution. In the Himalayan Buddhist tradition, rather than choose a life of renunciation, women are sometimes made renouncers by their parents if they come from families with more than one daughter. In some ethnically Tibetan communities of the Nepal Himalayas, as Shneiderman describes in this volume, second daughters are expected to become renouncers. This equation demonstrates a mutual material logic: young women who will not marry get to study without costing their families money, either for their education or for their dowry. By sending daughters to nunneries, poor families are less burdened by the economic cost of children, and girls get educated. By collectively valuing religion and religious practitioners, a community can advocate widespread education and literacy, as does the respected ascetic Chomo Khandru with whom Shneiderman worked in Nepal.

But communities support male institutions of renunciation much more consistently, and with much more money, than they do female institutions of renunciation. In this volume, Gutschow leads us through the dramatic disparity of communal support for monks and monasteries over that for nuns and nunneries in the Indian Himalayan region of Zangskar. To become a monk is to be assured of material assistance from lay villagers, but to become a nun may still require economic self-sufficiency. In this instance, and in many others throughout this volume, the ideology of transcending gender does not translate into everyday practice, and women, and women's institutions, must struggle to survive financially.

However, not all renouncers choose to affiliate with larger institutions, and this choice means more independence but less material security. Being a member of a communal institution is one way to relate to other renouncers and to the laity. A community of women is protective, from men, from naysayers, and from doubting, angry, or resentful householder community members who might use the logic that if a woman has rejected her place in the world, she can be treated with violence or disrespect, as if she is no longer truly a member of society. Communities of like-minded women can protect, as can ideologies of religious achievement that do not judge and even value displays of nonconformist or seemingly insane behavior. Conversely, isolation in a patriarchal setting may mean that women feel more exposed to social approbation and also to gender-based violence (cf. Khandelwal 1997). Ashrams, local communities, and monasteries can provide protection, but renunciation can also mean a heightened vulnerability.

Solitude, Isolation, and Mobility

The women renouncers with whom the contributors to this volume worked report several different kinds of tension in their personal and communal lives. One we heard often was the tension of choosing between an alternative community that offers both to sustain and protect women who have made radical life choices, on one hand, and the religious pressure to perform solitary religious practice, on the other. For some, institutions were just another form of hierarchical and subordinating social division. For others, group living offered legitimacy, security, and institutional support for religious discipline.[11] Despite this diversity of practice, renouncers in both the Hindu and Buddhist traditions are instructed at the textual level to practice alone, wander alone, and reject society of all kinds.

Isolation is both a test of strength for women and a rite of passage, a challenge as well as a danger. Some renouncers insist upon it, while others choose to renounce only within safe communal structures. For many women in South Asia, being alone is a frightening concept, particularly because it contrasts so strikingly with the intense sociality of family life, where solitude, equated with loneliness, is rarely sought. The solitude ostensibly required to renounce properly is both a mark of courage and strength and an indication of the antisocial ideology of textbook renunciation. For women renouncers themselves, isolation or solitude—physical or social—is sometimes talked about as an experience that can teach the conquest of fear of physical vulnerability or social reprobation, and sometimes as an occupational hazard.

Sometimes women who renounce do so precisely because they want to be free from suffocating social settings, and aloneness is possible under the rubric of renunciation. The Bengali Baul tradition Knight describes in this volume generally requires couples' *sadhana* (religious practice) rather than solitary *sadhana*, and yet the women Knight interviewed more often take renunciation alone. Their reinterpretation of appropriate religious practice seems to rely on the Hindu prescription of solo *sadhana*. The very possibility of solitude—or at least the possibility of being a woman free of the structure of marriage—encouraged their choices to renounce, even in ways that would be socially questioned in their own tradition.

One of the key practices of renunciation, at least on paper, is mobility, and indeed the freedom of movement is a large part of why many women choose to renounce. Women can use renunciation "to justify their mobility," as Knight puts it. Restriction on women's mobility in many parts of South Asia remains high, especially for women of lower castes (World Bank 2005). Women renouncers are publicly allowed to move between locations in ways that householder women are not, and a number of the women in this volume most likely renounced in order to move away from untenable situations—domestic abuse or alcoholism—in natal or marital homes (cf. Hausner, this volume). The need to leave home—demanding parents or parents-in-law, abusive husbands, bad treatment as widows, or confining social requirements—is one reason women may decide that householder life is not for them.

On the other hand, the pleasures of exploration and the opportunity to see the world can inspire renunciation. Indeed, for many renouncers in this volume, visiting holy places is one of the rewards of being itinerant, a way to articulate the human body against a sacred site in hopes of purification and attainment. Travel is popularly understood to be hard—pilgrimage, for example, is experienced as an ascetic endeavor for both renouncers and laypeople—and any community that voluntarily undertakes long or extended journeying is believed to be strong in some material or spiritual sense. Travel is a condition of hardship, and itinerancy a condition of insecurity. As for solitude, a public reinterpretation of a fearsome condition—perpetual travel—is part of the contract of renunciation.

At the same time, we cannot assume that all women renouncers spend their lives as wanderers. Not all renouncers want or choose to live a peripatetic life. After they initially leave home—to travel with a guru or to visit a pilgrimage place—most women decide to create an alternative home base, and create new relationships with the people in their new places. "Home" looks very different for a renouncer than it does for a householder woman, but some of the same kinds of duties and responsibilities may arise (Hausner and Khandelwal, both this volume). Renouncer women who have found new home bases may themselves become part of the attraction of a particular pilgrimage place. In the sacred city of Haridwar, Radha Giri stubbornly refused to move from a particular location she chose for its proximity to a holy pipal tree sprouting on the banks of the Ganges; the site itself had become holy in the local community because she and other renouncers before her had invested it with spiritual power (Hausner, this volume). Radha Giri argues that all the traveling required of Hindu renouncers is superfluous to her religious commitments; she would rather live a life of practice in a singular location.

The sense that religious pursuit means a renouncer is more intimately connected to holy places than an average pilgrim is a powerful theme in the lives of women renouncers. Shneiderman's mentor Chomo Khandru was initiated in the *gompa* (temple or monastery) of her home village when she was still a young girl. Like Radha Giri, Chomo Khandru talks about her days as a trader between the regions of Mustang and the border areas of Tibet as difficult in part because she would rather remain sedentary in her home village, Lubra. But she does acknowledge that her traveling commitments released her from sedentary domestic responsibilities.

Womanhood in South Asia: Body, Sexuality, Celibacy

Just as the lives of renouncers in real life look different from those of renouncers in religious texts, the experiences of women who renounce are different from those of women who do not (Denton 2004). The assumed life trajectory for South Asian women diverges considerably from the paths chosen by the women renouncers in this book. For example, if the Hindu women with whom we worked in North India, where patrilineal and patrilocal family structures are dominant, had lived in the way of their parents' choosing, they would have married a suitable boy in an arranged wedding, moved into their husbands' parents' homes and been answerable to their new mothers-, brothers-, fathers-, and sisters-in-law. They would be expected to produce a healthy child, ideally a son, within a year or two. In the early years of marriage, their sources of emotional support would be few, and their status would be at the very bottom of their extended households (cf. Raheja and Gold 1994; Ostor et al. 1982).[12]

Lay Baul, Jain, Buddhist, and South Asian Islamic traditions similarly emphasize marital and procreative activities, but none of the women in this book followed that path. As renouncers, they are women who have consciously chosen to leave their homes, in a region of the world where women are very rarely allowed to live on their own. Their new lives look very different from the lives they have left behind. They are worshipped in some contexts, and maligned in others. Some could not go home if they tried. Others defy desperate family requests to return. They follow their own life paths, even if it means being shunned, outcast, or totally alone.

Despite the religious, regional, and circumstantial variation of their lives, almost all the women with whom we worked in this volume have left their natal or marital homes, and have created a new life and identity for themselves. (The one renouncer in this book who lives near her father and maintains daily contact with him [Hanssen, this volume] nonetheless undertook a rite of initiation that included a vow to detach from members of her family.) Rather than remain with householder families—in body or in spirit—the women in this book chose to follow paths of religion. Some used religion to leave lives they were supposed to follow. Others made painful decisions to leave because they felt so powerfully drawn to religious life. Whether religious practice and duty was the reason these women chose to leave so-called householder lives, or a convenient option that allowed them new leeway in an otherwise prescribed life story, religious practice—in whichever religious medium they had been reared—became the vehicle of departure.

The Body

Renunciation is intended to bring about a state of emotional and embodied detachment in the practitioner, along with a recognition that the phenomenal world is transient and illusory. The extent to which the women discussed in this book are actually on their own varies, but all agree that their religious practice is solitary, and that their searches for God, and good, are independent of social connection. But here we have the tiresome, tenacious lingering of the body and its undeniable socialization. To live in a body—a woman's body, no less—means that some social residue will always remain, both subjectively and objectively. To live in a woman's body means others will always treat a woman renouncer as a woman. Sometimes this means they will be glorified, and sometimes it means they will be denigrated, but most women in this book, usually with some exasperation, talk about how their gender can never be fully left behind, no matter their actions or level of realization.

Most of the women on these pages talk about their bodies as if they were something other than themselves. Different religious traditions view the self-body dichotomy in different ways, meaning religious practice takes on a slightly different slant. Nonetheless, each of the major religions represented here—Hinduism, Buddhism, and Jainism—shares an insistence that the enlightened self is entirely ungendered, as well as a theological focus that the material (and therefore gendered or sexed) body is ultimately unreal or ephemeral.[13] The specific goal of the religious practice may vary, but in each of these traditions, practitioners strive to see through what they call the illusory nature of material existence. Power over the body, or the ability to penetrate the seemingly solid bounds of the material world, is also metaphorical for power over the world (cf. Hausner 2007), and women who can claim to subdue bodily passion, desire, craving, and fear are renouncers who will gain respect from the most doubting householders, in all traditions.

The very state of embodiment is the focus of religious practice of many renouncers. Whether they focus on the body's illusory nature in Advaita Vedanta Hindu philosophy or its inherent emptiness in Vajrayana Buddhist teachings or its ephemeral time on earth in Jain doctrine, the practitioners with whom we worked were acutely aware of the profound paradox of living in a human body and knowing it is nothing but empty form. People must have bodies to live their lives, of course, and the women we describe here were acutely aware of this dilemma. In response to a question about why Radha Giri situated herself on the banks of the river Ganges, for example, the *mataji* (or holy mother) asked Hausner back, "Where do you want me to live, the railway station?" Questions of embodied location are the exigencies of existence, she implies, not necessarily related to spiritual quest (Hausner 2007). Life is reliant on the body, but ultimately embodiment is irrelevant to spiritual achievement. How our subjects use their time in their bodies—how they divide up their days, or guard a balance between solitary religious practice and public religious oration, or choose between being alone and being identified with a particular community—tells us a lot about the religious traditions they come from, and the cultural priorities that each chooses to embrace or reject. As Vallely writes of a Jain nun, "Her body was not a tool for the well being of others, it was hers alone and therefore her decision whether to nurture or abandon it."

In one way, the women in this book regard bodies—their own and other women's—in the same way as scholars of embodiment do: as a product of social experience. Although they come from different traditions and follow different spiritual paths and practices, the women in these pages talk about their bodies as objects in order to distance themselves from the palate of social experience within which they live. Vallely tells us that "perhaps more than in any other way, Jain women narrate their own experiences with reference to the body," and that the " 'worldly' comes to be identified . . . in terms of their former socialized bodies, as well as . . . the socialized bodies of the lay women around them." Renouncer women experience what most anthropology students are taught: bodies carry the marks of the social world, daily experience, and human interaction.

If the body carries the weight of worldly social life, renunciation from the world becomes largely an effort to distance oneself from the body. Most women renouncers have struggled with the social expectations placed upon their women's bodies, as unmarried women, wives, or mothers. Indeed the assumption that they would use their women's bodies in the functions of social and familial procreation may well have driven many of the women in this book to renounce. In certain schools of the Jain tradition about which Vallely writes, training to detach and cease identification from the body is explicit and set out in a standard curriculum. In the Bon and Buddhist traditions about which Shneiderman and Gutschow write, religious liberation is impossible in a woman's body, and young nuns pray to be reborn as boys so they might quickly be released.

In the South Asian context, renunciation typically involves restrictions on eating, sleeping, and sexual activity, and this volume reveals the variety of ascetic practices within these traditions. For example, Jain religion elevates fasting with the aim of nonviolence as a central practice, a way of minimizing the effects of living in a body (Vallely, this volume). In describing the life of a Hindu woman renouncer, Khandelwal (this volume) emphasizes the renunciation of sleep in the interest of ritual and charitable activity. Vajrayana Buddhist texts similarly argue that practitioners should refrain from sleeping on a high or luxurious bed, cautioning that overindulging in sleep might lead to mental degradation and ignorance (Holmes and Holmes 1995:188).[14]

Celibacy is also embodied in that its meaning and practice have to do with sexual fluids and a particular ontology of the body. The danger of sexuality is not only that it creates certain kinds of emotional entanglements and social obligations (particularly in the event of pregnancy), but also that it involves an exchange of bodily substances. In South Asian theories of bodily fluids, substances are not merely symbolic of emotions and social relationships, but actually contain and transmit sentiment. Whether shared bodily fluids are understood as dangerous to purity, vitality, or pristine health, the meanings associated with celibacy—and the restraint of socially embedded and embodied activities more broadly—are simultaneously physical, emotional, social, and spiritual.

Sexuality and Celibacy

In most parts of the world, both sexuality and celibacy practices are gendered through discourses of female chastity and respectability. An unusual inversion to the theme of restraint and celibacy occurs in Baul tradition, about which Hanssen and Knight write, whereby ritual intercourse is reputed to be the most direct route to liberation. Tantric forms of renunciation do not necessarily require celibacy: Vaishnava Bauls valorize the act of a couple renouncing, rather than the individual choice to leave marriage and family for a life of celibacy. In contrast to dominant social paradigms, Baul women and men ritually use female bodily effluvia to gain power (Hanssen, this volume). Bodily secretions are linked with bodily power: Bauls seek to retain their sexual fluids through yogic techniques; women's menstrual blood, so polluting in Brahmanical Hindu doctrine, is central in their pursuit of spiritual enlightenment. In this tradition, a woman's body is a means to religious realization, not an obstacle to it.

Ironically, single Baul women who take renunciation alone rather than as part of a couple are marginal to the Baul ideal, which is itself subsidiary to the more elite forms of solitary and monastic renunciation in South Asia (Knight, this volume). Highlighting the strategic uses of celibacy, Knight shows how a single Baul woman may try to ensure social protection by associating with more prestigious—and celibate—Hindu *sannyasinis*. This identification normalizes her anomalous position as a single woman, offers protection for a presumed celibacy, and places her in a position to receive alms from Hindu householders. The dangers that single Baul women experience because they live outside the realms of procreative social experience are reflected in other traditions as well: Gutschow discusses village scandals involving rape in monastic institutions in Zangskar, and Shneiderman describes how the celibate woman Chomo Khandru carried a knife to protect herself from forced "marriage" during her trans-Himalayan trading journeys. Elsewhere, Khandelwal (2004, 1997) has discussed similar constraints faced by Hindu women living as solitary renouncers.

The chapters in this book demonstrate that female celibacy does not support patriarchy in any simple way. Vallely's work in this volume shows how lay Jain girls' socialization into modesty does not mean ignoring the body; rather, it means simultaneously concealing and revealing the body, a process she calls, citing Frigga Haug, a "sexualization of the female body." Vallely writes that "even the religious practices of lay women . . . do not remove them from their association with sexuality. Instead, religious observance is a fundamental marker of sexual purity and female honor." Female Jain renouncers thus cultivate an attitude of indifference to the well-being and appearance of their bodies, and they master the discourse of bodily objectification through fasting and the denial of bodily comforts and beauty. After a period of training at a monastic institution, their body language becomes desexualized, their longer stride and forceful conversation suggesting self-assurance rather than sexual restraint. They are contained in their comportment and respectful with their superiors, but their bodily restraint denotes the Jain attitude of nonviolence rather than the feminine modesty that signifies sexual respectability. Vallely's discussion illustrates the difference between social concerns with chastity and the practices of celibacy. Abstinence in the name of chastity sexualizes the female body according to patriarchal discourse, but religious celibacy is rather a reappropriation of power and strength.

Celibacy as Political

As with sexual practices and identities, celibacy can be a political act. Sobo and Bell note that monastic communities in various parts of the world may have adopted celibacy as a critique of particular social arrangements or social structures maintained in part by the sexual relations they have also constituted (2001:18). In a part of the world where sons are the primary security for aging women, most of whom own little property, a woman's vow of celibacy may be more about giving up children and family than it is about sex.[15] For most South Asian women, marriage is the only legitimate context for both sex and childbearing. In rejecting sex, female celibates shun marriage and a host of other social expectations. In doing so, they arouse awe, ire, or both. Bauls, however, appear to have it all, as couples may first have a child and then take renunciation in a context that allows continued sexual activity, albeit ritualistic.

Feminist scholarship has shown that pregnancy and childrearing has been promoted as a kind of service to the nation in particular historical and social contexts. In other circumstances, often with regard to minority populations, fertility has been constructed as a threat to the nation. In New Delhi in 2005, billboard signs announced a two-child campaign—"*hum do, humare do*," or "We Two, Our Two"— to promote contraceptive use as part of national service. In this volume, Menon shows us how female celibacy is put to nationalist use in her work on the political passion and militancy of Hindu women renouncers. Their celibacy is presumed to empower them politically and to lend religious authority to their political agenda. Refraining from sexual activity marks them as linked to the world of greater spiritual ideals rather than familial concerns; they are inspirational figures to be emulated in their strength, determination, and commitment to Hindu nationalism.

The celibacy of nationalist Hindu renouncers serves to motivate political commitment but is not generally held up as a practice that young women should follow. While male celibacy may be upheld as exemplary in Hindu disciplines of the body (cf. Alter 1992),[16] female celibacy is more likely to be set forth as an exception rather than a model to be emulated by ordinary South Asian women. Female celibacy is threatening for reasons of political economy: if the number of women choosing a celibate or renouncers' lifestyle were to increase, who would provide the reproductive labor for families? Female celibacy is equally threatening for discursive reasons, for it flies in the face of patriarchal assertions that women are unable to control their sexuality and are generally unfit for the solitary life of a renouncer. If women are allowed to make decisions about their sexuality—and their celibacy—how can they be properly controlled by male protectors?

The threat that female celibacy may pose to society can be seen in the example of the Brahma Kumari movement, which emerged during the 1930s in a merchant community in Sind, now part of Pakistan (Babb 1984). Founded by a wealthy businessman, women constituted its core membership; they were restricted by seclusion practices and a lack of education. The men in the community, on the other hand, were not only absent for long periods for business ventures abroad, but were reputed to have frequent extramarital affairs while away. The early teachings of the movement focused on the hypocrisy of marriage: while Hindu orthodoxy proclaimed that the husband was supposed to be a "god" to his wife, contemporary husbands were brutish and unworthy of worship. Unmarried women and young wives took vows of celibacy against the wishes of their families and sometimes without the knowledge of their absentee husbands. The movement faced ferocious opposition from society, and Babb points to celibacy as the reason it was so threatening: "In the Hindu world for a married or as yet unmarried woman to renounce her sexuality is to express a radical and unacceptable autonomy" (403). The complex cosmology of the Brahma Kumari movement held that in the pure period of human history, children could be conceived by yogic powers; it was the advent of sexual intercourse that marked the transition to a hellish human experience and the oppression of womankind (406). The critique of the family articulated by the Brahma Kumari movement was a political statement, and the promotion of celibacy was crucial to that critique.

Celibacy can also be a tool for individual resistance, particularly in the face of family pressure to marry. In South Asian cultures in which celibacy and virginity are highly valued ideals, parents face a dilemma when daughters insist on pursuing celibacy rather than marriage. All of the essays in this volume attest to the fact that "[t]he politics of gender are deeply implicated in the practice of celibacy" (Sobo and Bell 2001:19). If the pursuit of pleasure and sexual fulfillment can be empowering for some women confined by ideologies of chastity, abstinence can be similarly empowering for others. South Asian women articulate a variety of critiques of conjugality: in Rajasthani songs, for example, women sing about their marriages, arguing that husbands generally don't live up to their own stated ideals (Raheja and Gold 1994). The Brahma Kumaris went further to say that marriage itself needs to be reconfigured. In this volume, celibate women's lives exemplify the even more radical statement that marriage is not necessary.[17]

Understanding Celibacy within Sexuality Studies

The burgeoning field of sexuality studies holds gender analysis as central, but a feminist perspective can less readily be assumed.[18] Still, feminist scholars working outside Euro-American contexts—often anthropologists—are increasingly drawing attention to the culturally embedded assumptions of much theorizing on sexuality. For example, sexuality studies currently offers little room for talking about abstinence and celibacy as anything other than pathology or repression. Much work in sexuality studies seems to take as its starting point two intertwined notions: first, that sexual pleasure is both morally good and healthy, and second, that sexuality (whether heterosexual, homosexual, or queer) is central to identity. Understanding South Asian renunciant traditions requires suspending both assumptions.

The terms celibacy, abstinence, and chastity are often used interchangeably, but we believe that they are distinct phenomena and that each is gendered in particular ways (cf. Khandelwal 2001). We use celibacy to mean a purposeful state of complete sexual abstinence, usually permanent, and part of a lifestyle that may also include some or all of the following: restrictions on food and sleep; prohibition of intoxicating substances; avoidance of entertainments; renunciation of wealth, social status, and kinship ties; and engagement with ritual practice. The ideal celibate in South Asian religions (and possibly all religions) is generally male. This is especially true in textual and ecclesiastical perspectives, but it is also often the case in ethnographic research. Hindu celibacy, for example, is usually seen in physiological terms as a male pursuit intended to retain seminal fluids within the body (Alter 1992, 1994, 1997; van der Veer 1989). The feminist scholars whose work appears here document the specificity of women's voices in celibate renunciation.

What of the term "chastity"? Conventionally speaking, chastity is the proper female counterpart to the male pursuit of lifelong celibacy. It requires temporary or intermittent periods of abstinence that are compatible with marriage institutions.[19] Chastity, which more often describes ideal femininity than ideal masculinity, is a properly regulated expression of sexuality, imbued with status and morality, supportive of marriage ideals. As such, chastity must be distinguished from the lifelong pursuit of (ideally male) celibacy, which is typically associated with an unmarried state (Khandelwal 2001:163). This book explicitly engages the discourse of celibacy—even noncelibate Baul traditions enter into a dialogue with the dominant renunciant traditions that place celibacy at the center of practice—but notions of female chastity and respectability nonetheless appear frequently in the narratives of women with whom we worked. This repeated emphasis reflects the popular concern with women's behavior that usually crops up in the domain of sexuality.

Sobo and Bell (2001) note that the failure to include celibacy in the study of sexuality results from the erroneous assumption that celibacy is synonymous with asexuality, a simple lack of sexual desire. On the contrary, they argue, asexuality is not coterminous with celibacy, unless it has been carefully cultivated by celibate practices. Celibacy assumes the presence of desire, for, as the authors point out, those who are not attributed with sexual desire within a particular culture cannot be classed as celibate. In cultures where heterosexuality represents a hegemonic ideal and same-sex relations are negatively sanctioned, for example, members of the heterosexual mainstream can scarcely imagine homosexual celibacy. Of course, in many contexts, and depending on the concept of same-sex desire at work, homosexual celibacy is indeed possible.[20] We are learning that celibacy is not asexuality but rather a particular form of sexuality, a conscious set or series of practices and behaviors.[21] If in Jain doctrine, the physical world is a "moral theater," it is perfectly logical that young aspirants would train their bodies to project a celibacy that bespeaks strength and discipline, not shyness and modesty (Vallely, this volume).

South Asian traditions of celibacy reject conjugal sexuality outright. Moreover, like the Euro-American prototype of the Catholic monk, they prohibit any kind of genital stimulation that results in arousal. Further, in Buddhist, Hindu, Jain, and Baul celibacy, a stance of emotional detachment is carefully cultivated. Although solitary pursuit may mean that women renouncers are more vulnerable to male sexual aggression, the sexuality that women renouncers spoke about most often with us came in the form of emotion. Suppressing or transforming affect became part of their religious practice. Emotion and intimacy are core issues for women renouncers, and every woman in this book has had to grapple with these manifestations of the body, as they might term them (Hausner 2007; Khandelwal 2001).

Clearly, how we understand "sex" in a given context determines how we understand celibate subcultures: what counts as celibacy depends on what counts as sex. Taking celibacy seriously as a goal, lifestyle, or philosophy gives us a new point from which to critique assumptions of sexual expression and fulfillment as central to identity, health, and self-development. This volume offers a careful consideration of the lifestyles of celibacy, and what they mean for women who pursue them.

South Asian Renunciation in Global Perspective

The research and writing for this book were completed at the turn of the twenty-first century, when global processes are on everyone's mind—anthropologists and scholars of religion are no different from political scientists and economists in this regard. In the last two decades, scholars from various disciplinary backgrounds have warned against the tendencies to take "the nation" as primary unit of analysis, to assume culture is tied to place, and to overemphasize the boundedness of "culture areas" such as South Asia. Recent theorizing has argued persuasively for a focus on transnational rather than international models, and for tracing historical linkages between places rather than comparing them (Kaplan and Grewal 2002; Gupta and Ferguson 1992). It is no longer theoretically justifiable to think of local cultures as isolated from the rest of the world, and mounting historical research suggests that, empirically speaking, it never was (Pollock 2003; E.R. Wolf 1982).

Recent technological changes have had massive social implications. At the beginning of the twenty-first century, more people travel with greater frequency across the borders of nation-states, and new kinds of communities have become imaginable in the media of cyberspace. Renouncers travel far from South Asia as teachers, and their audiences include both members of South Asian diasporas and so-called Westerners. North Americans, Europeans, and Australians—and increasingly people from Japan, Taiwan, and other Asian, African, and South American countries—in turn travel to South Asia to learn meditation and yoga from monks, nuns, and swamis; a few become "Western renouncers."[22] Of these, some remain in South Asia more or less permanently, while others return to their homeland to live in monastic institutions. At the time this volume was written, the Pluralism Project reports that a small community in Lincoln, Vermont, is breaking ground for the first Tibetan Buddhist Nunnery in North America.[23]

Hinduism has become globalized through the guru path, in which women ascetics have been particularly prominent since the 1970s. Karen Pechilis notes that Hindu female gurus, virtually all of whom are considered ascetics, often visit or reside in the United States; she traces this development to the 1965 abolition of immigration quotas and to popular interest in Asian religious traditions in general and meditation in particular (2004:10,34), which flourished in Britain and the United States in the 1960s among a generation of hippies.[24] In the West, gurus tend to accept participants from all ethnic and religious backgrounds, to emphasize a congregational mode of worship, and to disassociate their gatherings from organized religion (36).[25] Even the Terapanthi Jain nuns travel: the Pluralism Project reports the attendance of five *samanis*, or semi-ordained nuns of the Nav Terapanth Jain order, as the special guests of a 2004 convention of the Young Jains of American in California.[26] Religious objects and rituals, too, have spread outward from their original settings into new contexts and places. The practices and symbols of South Asian ascetic traditions—meditation, yoga, fasting, *vipassana* retreats, and *malas*, or rosaries—are now the practices, icons, and commodities of a global spirituality (Moran 2004).

While international travel is not unusual for today's renouncers, the women presented in this book are not global figures in this most obvious sense. Baiji and Radha Giri are not "Mahagurus" who have mass appeal or mass followings (Khandelwal and Hausner, both this volume). While Baiji is a guru, she has not traveled outside India and does not have foreign disciples, although many of her followers live or travel abroad and tell her about their experiences. Radha Giri refuses to move from her spot on the Ganges river, but she does have both local and foreign devotees, who sponsor her with a little money. The women portrayed in this book remind us that not all people who live in this globalized and networked world speak English, have access to email, or travel great distances. At the same time, that these women do not have websites or international experience does not mean that their lives are unaffected by global processes.

All of the Hindu, Buddhist, Jain, and Baul renouncers portrayed here either articulate or resist modern discourses on ascetic traditions. For example, Hinduism emerged on the global stage during the colonial period, and the practice of *sannyasa* transformed itself during the late nineteenth century in response to both external critiques and internal reforms, arriving at a new emphasis on activism in service of the Indian nation. Khandelwal (this volume) describes the service-oriented modern renunciation of Baiji, and Menon (this volume) highlights the Bhagavad Gita as the text through which a *seva-* or service-oriented renunciation is elevated and refined specifically toward the establishment of Hindu nation in India.[27] Alternatively, Chomo Khandru explicitly critiques a modernizing discourse that privileges elite monasticism and Tibetan textual training imposed by well-intentioned outsiders and foreigners who wish to support "tradition" (Shneiderman, this volume).

The contributors to this volume specifically examine the historical and transnational constructions of "tradition" itself, rather than accept at face value the rhetoric of religious practices as following ancient origins. We also try to understand the way in which current renunciant practices, meanings, and institutions within South Asia, including those that exist far from urban centers, are transnationally constructed.[28] Shneiderman argues not only that fluid religious identities in Lubra, Nepal, have given way to more rigid and narrowly defined ones, but that this can only be understood as the influence of extra-local entities such as the Tibetan monastic establishment in exile, the Nepali state, and foreign development organizations. Monasticism is too often viewed as categorically "traditional," but Shneiderman demonstrates that "traditional" ideas and institutions in remote parts of the Himalayas are dramatically constituted by processes that originate in other parts of the world and are sometimes the result of Euro-American visions of what constitutes Tibetan religion. The resultant shift toward more formal monastic training has not meant greater opportunities for women to pursue renunciation in this remote area of Nepal. Chomo Khandru, like many other renouncers, bemoans the corrupting influence of market economic relations—"the dharma of money"—on monastic life.

The field of South Asian studies has increasingly come to include research on diaspora and other phenomena that are not contained by the geopolitical or supposed cultural boundaries of the region. To the extent that South Asian religious renunciation is understood through a particular body of literature in anthropology, sociology, and religious studies, the issues with which this volume grapples are squarely situated in area studies. Yet renunciation, like all religious discourse and practice today, is constituted by processes that exist beyond the geographical space of South Asia. The work in this book suggests that it is possible to enrich the scholarship in area studies while also doing ethnography across borders.

Knight (this volume) explores a renunciant Baul tradition that challenges conventional understandings of South Asian renunciation in multiple ways: with specific reference to cross-border issues, it straddles both the national boundaries of India and Bangladesh and the categories of Hindu and Muslim that are too often monolithically mapped onto those countries, particularly in this era of religious nationalism. Knight's work elsewhere shows the influence of foreign appreciation of Baul music on Baul women and men, which can lead to foreign travel and even foreign marriage (Knight 2005). Her research illuminates cultural processes that supersede the nation-state at the same time that it enriches the field of South Asian studies. Throughout this book, we emphasize research that incorporates an awareness of how global processes impact women renouncers and affect ways of renunciation in particular localities.

Participant Observation in Ashrams, Nunneries, and Tents

All the research in this volume is ethnographically grounded. The authors learned South Asian languages and spent extended periods of time with the women about whom they write. What does it mean to sit by the *dhuni*, or sacred fire-pit, all day with notebook in hand? What insight is gained by standing at a kitchen table alongside disciples, faced with a mound of pumpkins that need chopping, and armed with a kitchen knife rather than pen and notebook? What epiphanies occur when the researcher performs publicly with Baul singers rather than observing from the sidelines? These authors demonstrate that participant observation can yield fascinating insights not available through scriptural study or even formal interviews.

"Ethnography" is a trendy word in the academy these days, and it can include a wide variety of methodological tools. Ethnographic research as we mean it here is not simply a matter of enriching one's thesis with data gleaned from interviews with real people. Rather, the art of listening cultivated in ethnography prompts reflection on the assumptions that are always embedded in research questions. In line with classical anthropological methodology, ethnography aims to interpret culture across various domains. Thus, the authors not only observed women renouncers in multiple contexts and tried to understand a variety of perspectives, but they also juxtaposed the women's verbal statements with their actual behavior. For example, Gutschow follows Yeshe from the nunnery where she is responsible for certain chores to the home of her mother who also needs her assistance; she then juxtaposes these two different domains of social life to a classical ideal of monastic life on the one hand and to the actual daily lives of monastics on the other. What emerges is a sense of the multiplicity of discourses and domains—ritual, domestic, monastic—as well as the complexity of social constraints with which a Tibetan Buddhist nun in Zangskar must contend.

The research methodology of participating in people's everyday lives, listening carefully, reading between the lines of formal answers to formal questions, and thinking across domains can yield striking insights. Moreover, an empirically grounded analysis helps to deal with the dynamic interplay between ideas and events, symbols and social realities (cf. Sobo and Bell 2001). Menon's chapter demonstrates that when the primary subject of one's inquiry is a public figure with an international reputation, certain limits on interpersonal interaction are imposed. Still, Menon examines the discourses of these powerful ascetic women in the social milieu in which they are delivered. These essays illustrate that ethnographic research is a particular kind of social relationship.[29] In most cases, the relationship is an intimate one.

An Ethnographic Perspective

A particular challenge facing ethnographers working on renunciation is that cultural ideals and religious propriety sometimes caution renouncers against talking about their lives prior to initiation, for this is the life they have deliberately left behind. It is bad form for Hindu renouncers to speak freely and openly about their childhoods, families, or troubles before renunciation. Conversely, in Vajrayana Buddhism, a tradition of spiritual biography leaves open a possibility for women to recount their histories. Some of our authors had to fill in the gaps about renouncers' backgrounds or ponder upon possible interpretations of their silences. Are particular stories too painful to tell? Are certain biographical details embarrassing? Might they suggest that their renunciation was an act of desperation rather than a spiritual calling? Must some experiences remain hidden from those who are not capable of understanding? Might information about their good fortune or accomplishments be interpreted as self-aggrandizement? Or is it possible that talking about oneself should be avoided simply because it will cultivate the ego? Just as there are limits to using scripture as a primary source—we will never be able to answer certain questions about the sociological significance of the debate between Maitreyi and Yajnavalkya as reported in the Brhradyaranyaka Upanishad—so too are there limits to ethnographic data. These limitations are intensified by the goal of deconstructing the self in renunciant traditions.

All of these women follow renunciant traditions that have scriptural referents, which usually encapsulate an elite male perspective. For this reason, an ethnographic approach can offer insight into women's perspectives on—and experiences of—textually-defined monastic traditions. We do not seek to substitute ethnographic methods for textually-based ones, for the two are complementary. Thus, this work complements that of feminist scholars working with textual sources who have written on women ascetics,[30] just as it complements that of feminist scholars working ethnographically with householder women in South Asia.[31] Ethnography allows us to access closely and personally the everyday enactment of theological and philosophical ideas. It allows intimacy with religious practice and experience not accessible through studying texts.

Paying attention to the subjective life of women renouncers further demonstrates that their thoughts and actions are not exclusively determined by patriarchal models of renunciation. It also shows that women's behavior does not always reflect their subjectivity. Just as women may veil for a variety of reasons (Fernea and Fernea 1997; Raheja and Gold 1994:xxii), when women renouncers take on male attributes or behave in accordance with norms for femininity, they may do so either with respect for social convention or with cynicism. While male biographers are fascinated by accounts of medieval Christian women saints' cross-dressing, attributing to them numerous symbolic interpretations, the choice to don male clothing on the part of the women themselves was more likely a pragmatic strategy to travel safely or escape marriage (Bynum 1991:38). In a different context, Khandelwal (this volume) eventually comes to learn that when Baiji engages in ritual activity in her Haridwar ashram, she does so self-consciously and with a particular didactic intent. Women renouncers are humanized when one is attentive to their struggles, compromises, and reflective musings. This volume is designed as a way to bring the experiences and perspectives of women renouncers in South Asia to life, even in contexts very far from their own.

Conclusion: Grounding Women's Renunciation

In the context of modernity, it is easy to assume that the number of renouncers in South Asia might be diminishing. But in the small and sparsely populated Himalayan region of Zangskar located in the southeastern corner of Indian Jammu and Kashmir, it is the number of monks not the number of nuns that appears to have leveled off, due to increased opportunities for salaried jobs and new forms of wealth and prestige in a growing cash economy. Because of less access to government and military employment, women who seek opportunities for education, service, and travel see the nunnery as a desirable career option. This has resulted in a rising number of applicants at existing nunneries and construction of many new nunneries in the region (Gutschow, this volume). In very different social and economic circumstances, Jain nuns are today more numerous than their male counterparts (Vallely 2002). Although it is notoriously difficult to obtain statistical information on Hindu renouncers, Ramaswamy (1997) and U. King (1984) both suggest that renunciation has become more accessible to women, not less, over the last century. Previously all-male monastic institutions have begun to admit women, women-only ashrams have been founded, and independent women gurus are prominent in both South Asia and on the global stage.

While women are often respected for their choice to renounce, the path of renunciation can be tremendously difficult. Some of the renouncers in this book told their interviewers that they would prefer their daughters marry than live as they had, or at least choose to remain in society even if they would rather live with their gurus (Knight, this volume). Religion is a communally agreed-upon value in the places these women live, so the extent to which they reject or defy social norms—or embrace them in new forms—varies among the subjects of this book. Sometimes the women are looked down upon or pitied—some women's ashrams are truly homes for the aged and destitute—and some are revered. All the women in this book interact with lay society as well as the community of renouncers, but many are shielded by—and even make use of—the public view that renouncers lie outside rigid cultural norms. Sometimes this perception leads to a kind of respect for other-worldly powers or highly unusual behavior; sometimes it leads rather to a kind of public disgrace. The tension between public support and public approbation is ongoing in the lives of most renouncers.

A number of the women in this book have attained a considerable amount of social and even political power (Menon, this volume), which is sometimes derived from their status on the margins of society (Hausner, this volume).

If they can tolerate the challenge of renunciation, subvert the logic of sexuality, or gain control over the clamoring desires of their bodies, the logic goes, they must be superhuman beings, worthy of the popular devotion due the most spiritually advanced practitioners, regardless of gender. In one sense, the religious values of South Asian traditions clearly place the phenomenal world below the spiritual one, and a woman who can defiantly choose the latter, risking the reprobation of the former, eventually deserves its respect. But she must prove herself as a real renouncer and, in some cases, as a woman able to withstand the repeated challenges of social disdain.

All the women in this book struggle with the divergent requirements of social and spiritual power, or "conflicting sets of social and spiritual expectations which female practitioners face," as Shneiderman puts it. The Himalayan Bon tradition to which Chomo Khandru belonged explicitly emphasized a "practical" spirituality, which allowed more women to participate from the outset. A.K. Ramanujan (1982) has argued that male saints are often attributed with feminine qualities, because women's "natural" inclinations toward nurturing befit the role of a generous and committed spiritually powerful presence (see also Denton 2004). In this volume, Shneiderman cites Ramble to point out that celibate nuns in Nepal's Mustang region are addressed with male honorifics. Of course, as Khandelwal points out (1997), the rhetoric of religious communities does not always translate into material reality, and often women renouncers are faced with language that professes to ignore their female bodies right alongside treatment that clearly denigrates them as women.

Taking on the role of woman renouncer confounds the relationship between skill, gender, body, and practice in ways that challenge the practitioner and her community alike. The ways female renouncers participate in local communities can heighten a sense of their power, and the social and spiritual aspects of renunciant life are often intertwined in unexpected ways (Menon, Shneiderman, Khandelwal, and Hausner, all this volume). A female renouncer can even be dangerous, but not because of her sexuality: she is no longer constrained by assumptions that her primary social role is that of mother, and so she removes herself from the realm of procreation, freeing up her power to use as she will. At the same time, communities of women renouncers can offer an intimacy that looks very different from lay or householder social, physical, or sexual intimacy (Gutschow, this volume). This book reminds us that both sexuality and celibacy involve a rich variety of ideas and practices, even within the specific context of South Asian religions.

The women in this volume decided that the life their parents, husbands, or communities planned for them did not coincide with their own personal or religious aspirations. They were uninterested in the cultural and familial proscriptions that were laid out by the structures into which they were born or married, and which no one imagined that they would reject. They are gathered here in one volume despite the variability of region and religious tradition. Collectively, their stories show us something about how women live in South Asia through the lens of renunciation—that is, not always as wives and mothers. They also show us something about how renouncers live through the lens of gender—that is, with particular constraints and opportunities, because they are women.

Women's renunciant practices in South Asia are multiple, fluidly linked to embodied practice. Baul women in West Bengal and Bangladesh may be Hindu or Muslim, but they equally rely on the power of song as a mode of devotion. Women in India and Nepal—following diverse Hindu and Bon traditions—might similarly choose not to marry or bear children of their own, but rather support others' children in the communities in which they live. Most importantly, conscious reflection on the experience of embodiment links women in all these traditions, from Jainism to Buddhism, more than it likens them to women of the same tradition who have not renounced. Renunciation looks different when we consider women's stories, and South Asian women's lives look different when we consider the role of religious practitioner. The goal of this volume is to explore these dynamics, analyzing the ways questions of agency, sexuality, domesticity, and community change when we combine these two visions.

Notes

1. South Asia is an enormously diverse region, and we do not cover renunciation in Pakistan, Bhutan, Sri Lanka, or South India in this volume. Nor do we describe the lives of Christian or Theravada Buddhist nuns, both important contemporary traditions of renunciation for women. The goal of this book is not to produce a comprehensive catalog of women's renunciant traditions, but to offer detailed analyses of a small number of women's stories and circumstances, in the hope of inspiring ongoing thought about the lives of women renouncers in the region.
2. The work of Ann Grodzins Gold has been instrumental in establishing this approach as a way of understanding women's lives in South Asia. We thank her for her comments on this introduction in particular, and for her support and guidance in our work on these subjects more broadly.
3. We thank Anne Vallely for clarity on this point and on Jainism throughout this chapter.
4. There is some debate on this point; the one agreed-upon exception to the dominance of men and the male gaze in South Asian religious textual history is in the case of tantric texts, where women are espoused as necessary partners, and possibly as solo practitioners themselves (cf. White 1996, 2003; Shaw 1994).

5. Scholars disagree about the origin of renunciation in India: some believe that it grew out of caste hierarchy to institutionalize Brahmanical ritual power, while others believe that it grew out of a public opposition to such hierarchy (cf. Olivelle 1992; Thapar 1979; Heesterman 1964, 1993).
6. Although we do not have comprehensive data, our respective fieldwork leads us to believe that *sannyasis* and *sannyasinis* (institutional Hindu renouncers) who live in prestigious monastic ashrams usually come from high-caste backgrounds (also see Denton 2004), while street-dwelling *babas* usually come from low-caste backgrounds. This is an important (if difficult) area for further research (see Hausner 2002). "Lower-caste" ascetics were not permitted to join traditional high-caste regiments until the sixteenth century, when they were actively recruited as mercenary troops (Pinch 1996).
7. Our emphasis in this volume that renouncers' lives dramatically differ from householders' lives follows the work of Louis Dumont (cf. especially Dumont 1980[1966]), who uses this relationship as a base from which to understand Indian society. Dumont preferred to see these categories as entirely distinct, however, while we insist on analyzing the different ways renouncers engage with householders, or the laity more broadly, believing that these relationships can tell us a great deal about how renouncers actually live and function in South Asian societies.
8. In addition, some members of the laity support renouncers because they fear the consequences of denying a powerful yogic request.
9. For example, see Bouillier (1998) for a description of the relation between Nath ascetics and the monarchy in Nepal. On the other hand, Jain ascetics would not themselves call their interactions "exchanges," which would implicate them in worldliness, and thereby violence (Vallely, this volume). Even in this case, however, the monastic institution relies on financial support from the laity.
10. There are also popular discourses, however, that emphasize the corruption or the nonproductivity of renouncers, or that present them as an obstacle to modernity. Hindu sadhus are often depicted as a threat to the social order, for example: children are sometimes told that if they are bad, they will be sent to live with the sadhus.
11. See also, for example, the Kerala monastics studied by Sinclair-Brull (1997).
12. For similar structures in South India, see Wadley (1977) and Trawick (1990); in Bengal see Inden and Nicholas's classic study of kinship (1977); for Hindu women in Nepal see Bennett (1983) and Cameron (1998). Matrilineal kinship structures are also found in some communities in South India and on the Tibetan Plateau. There is fraternal polyandry in Zangskar, where Gutschow worked (this volume, 2004), and this was also practiced in Mustang until recently (Schuler 1987). Of course, in Tibetan Buddhist and Jain communities, there are cultural proscriptions for monasticism, but this does not detract from nuns' status as outside the standard social orders of procreation.
13. See Khandelwal (1997) and Leslie (1991) on the *atman* (soul) as ungendered in Hindu theological discourse.
14. Thanks to Hetty MacLise for bringing our attention to this text. Also see Mackenzie (1998) for a biography of Tenzin Palmo, a British-Buddhist nun who slept in a square box for 12 years on retreat.

15. Gold has shown that in Rajasthani oral tales, which offer a male perspective on renunciation, it is more painful to give up property and sons (either actual sons or the possibility of sons) than sex (1992).
16. Alter's ethnography (1992, 1994) reveals a similarly nationalist impulse that underlies North Indian wrestlers' efforts to promote celibacy, and thus vitality and masculine virility, throughout (male) Hindu society.
17. On the question of whether these critiques of marriage and husbands should be deemed feminist, Uma Narayan (1997) makes an important distinction. Feminist critiques of the institution of marriage may echo some of those voiced by nonfeminist women who nonetheless observe the suffering women face in their conjugal relationships, but a critical awareness of the gender dynamics within one's family or one's "culture" is not enough to make women feminists (11). The difference is that feminist critical analyses of institutions and practices usually "point to the *systemic* and *systematic* nature of the problems they focus on" (12).
18. Our discussion of sexuality studies benefited tremendously from the insights of Ellen Lewin and Harish Naraindas.
19. Thus, by the terminology employed here, most male and female Christian youth in the United States who call themselves "celibate" have, by the terminology employed here, taken vows of chastity or abstinence, for they envision a goal of temporary, premarital abstinence (virginity) as a prelude to conjugal sexuality and in the interest of romantic love and procreation. Chastity can also mean postmarital restrictions on sexual activity, which of course vary cross-culturally.
20. Foucault suggests the importance of abstinence or sexual austerity not only for Christians but also for the Greeks, whose moral concern with pleasure focused not on the particular mode of sexual practice but on its intensity (1985:44). The Greeks characterized a man by whether he demonstrated moderation, not by whether he was involved with women or boys, as in the example of Agesilaus "who carried moderation to the point that he refused to kiss the young man that he loved" (44). Homoerotic abstinence is also considered possible in many contemporary Christian churches that recognize gay or lesbian clergy. In the Catholic Church, for example, a gay priest is assumed to have a gay identity without any corresponding sexual expression, although recent scandals involving abuse of boys by Catholic priests in the United States has spurred a vociferous debate about the ordination of gay men.
21. This is further illustrated by abstinence trends in other contexts, such as among Christian youth in the United States, for example, which encompass a range of meanings and practices. Anthropologist De Munck reports that some of the young people he interviewed at the University of New Hampshire have vowed lifelong celibacy and refer to the Euro-American prototype of the monk who abstains from any kind of physical involvement for "the greater good" of serving God and humanity; most students, however, pursue a kind of premarital abstinence that prohibits penile-vaginal penetration but might allow for "kissing and fooling around a little bit" (2001:217). In the latter cases, abstinence is pursued in the interest of conjugal romantic love.
22. See Strauss 2005; Moran 2004; Bharati 1961.

23. Initially four women will live a monastic life under the tutelage of Khenmo Drolma, formerly Gina Kelley, an ordained Buddhist nun. See Joshi (2005) and http://www.addisonindependent.com/News/080105nunnery.html. Also see Simmer-Brown 2001; Mackenzie 1998; Chodron 1991; Allione 1984; for writings on and by American and European Buddhist nuns.
24. See Neill (1970) for a quirky but suggestive discussion of parallels between sadhus and hippies.
25. Tracing changes in American representations of Hindu sadhus in the nineteenth and twentieth centuries, Kirin Narayan notes that, despite increasing nationalist interpretations of Hinduism, the Hinduism associated with traveling gurus has a hybrid character detached from "India" as a place (1993:496).
26. See http://www.pluralism.org/research/profiles/display.php?profile=73776
27. Hindu nationalist endeavors also rely heavily on modern transnational processes for funding, leadership, and grassroots support (cf. McKean 1996; van der Veer 1988). Dirks has recently noted that communal tensions in South Asia and by extension among South Asians in the United States also reflect an increasingly transnational form of nationalism (2003:47).
28. Scholars disagree on to what extent this is a new or contemporary phenomenon. Our view is that this is an empirical question that must be answered in the context of specific topics and time periods.
29. Like all social relationships, the ethnographic encounter is imbued with power. Many feminist anthropologists have written about the politics of ethnographic research and writing (cf. Behar and Gordon 1995; Visweswaran 1994; and M. Wolf 1992). The chapters in this volume illustrate the multiple and complex axes of power at work when women anthropologists based in the U.S. or Europe interview women renouncers in South Asia. In some encounters, the anthropologist is clearly the more empowered party; in others, the situation is more complex.
30. See Denton 2004; Simmer-Brown 2001; Klein 1995; S. Young 1994; Shaw 1994; Wilson 1996; Ramaswamy 1992, 1996, 1997; K. Young 1987; and Findly 1985.
31. See note 12 above.

Bibliography

Allione, Tsultrim. 1984. *Women of Wisdom*. London: Routledge and Kegan Paul.

Alter, Joseph. 1992. *The Wrestler's Body: Identity and Ideology in North India*. Berkeley: University of California Press.

———. 1994. Celibacy, Sexuality, and the Transformation of Gender into Nationalism in North India. *Journal of Asian Studies* 53(1):45–66.

———. 1997. Seminal Truth: A Modern Science of Male Celibacy in North India. *Medical Anthropology Quarterly* 11(3):275–298.

Appadurai, Arjun. 1991. Global Ethnoscapes: Notes and Queries for a Transnational Anthropology. In *Recapturing Anthropology: Working in the Present*. Richard G. Fox, ed., pp. 191–210. Santa Fe: School of American Research Press.

Babb, Lawrence. 1984. Indigenous Feminism in a Modern Hindu Sect. *Signs: Journal of Women in Culture and Society* 9(3):399–416.

———. 1985. *Redemptive Encounters: Three Modern Styles in the Hindu Tradition.* Delhi: Oxford University Press.

———. 1998. *Ascetics and Kings in a Jain Ritual Culture.* Delhi: Motilal Banarsidass.

Behar, Ruth, and Deborah A. Gordon, eds. 1995. *Women Writing Culture.* Berkeley: University of California Press.

Bennett, Lynn. 1983. *Dangerous Wives and Sacred Sisters: Social and Symbolic Roles of High-Caste Women in Nepal.* New York: Columbia University Press.

Bharati, Agehananda. 1961. *The Ochre Robe.* London: George Allen and Unwin.

Bouillier, Veronique. 1998. The Royal Gift to the Ascetics: The Case of the Caughera Yogi Monastery. *Studies in Nepali History and Society* 3(2):213–238. Kathmandu: Martin Chautari.

Bourdieu, Pierre. 1977. *Outline of a Theory of Practice.* Richard Nice, trans. Stanford: Stanford University Press.

Burghart, Richard. 1983a. Renunciation in the Religious Traditions of South Asia. *Man* (n.s.) 18:635–653.

———. 1983b. Wandering Ascetics of the Ramanandi Sect. *History of Religions* 22:361–380.

Butler, Judith. 1993. *Bodies that Matter: On the Discursive Limits of "Sex."* New York: Routledge.

Bynum, Caroline. 1991. *Fragmentation and Redemption: Essays on Gender and the Human Body in Medieval Religion.* New York: Zone Books.

Cameron, Mary. 1998. *On the Edge of the Auspicious: Gender and Caste in Nepal.* Urbana: University of Illinois Press.

Chodron, Pema. 1991. *The Wisdom of No Escape.* Boston: Shambhala.

Coakley, Sarah, ed. 2000. *Religion and the Body.* Cambridge: Cambridge University Press.

Csordas, Thomas, ed. 1994. *Embodiment and Experience: The Existential Ground of Culture and Self.* Cambridge: Cambridge University Press.

Das, Veena. 1982[1977]. *Structure and Cognition: Aspects of Hindu Caste and Ritual.* 2nd edition. Delhi: Oxford University Press.

———. 1984. Paradigms of Body Symbolism: An Analysis of Selected Themes in Hindu Culture. In *Indian Religion.* Richard Burghart and Audrey Cantlie, eds. London: Curzon Press.

De Munck, Victor. 2000. Cultural Schemas of Celibacy. In *Celibacy, Culture, and Society: The Anthropology of Sexual Abstinence.* Elisa J. Sobo and Sandra Bell, eds., pp. 214–228. Madison: University of Wisconsin Press.

Denton, Lynn T. 1991. Varieties of Hindu Female Asceticism. In *Roles and Rituals for Hindu Women.* J. Leslie, ed., pp. 211–231. Rutherford, NJ: Fairleigh Dickinson University Press.

———. 2004. *Female Ascetics in Hinduism.* Albany: State University of New York Press.

Dimock, Edward C. 1966. *The Place of the Hidden Moon: Erotic Mysticism in the Vaisnava-Sahajiya Cult of Bengal.* Chicago: University of Chicago Press.

Dirks, Nicolas. 1997. The Policing of Tradition: Colonialism and Anthropology in Southern India. *Comparative Studies in Society and History* 39(1): 182–212.

———. 2003. *South Asian Studies: Futures Past.* UCIAS Edited Volume 3 The Politics of Knowledge: Area Studies and the Disciplines. Year 2003, article 9.

Doniger, Wendy, with Brian K. Smith. 1991. *The Laws of Manu.* New York: Penguin Books.

Dresch, Paul, Wendy James, and David Parkin, eds. 2001. *Methodology and History in Anthropology.* Anthropologists in a Wider World: Essays on Field Research, vol. 7. New York and Oxford: Berghahn.

Dumont, Louis. 1980[1966]. *Homo Hierarchicus: The Caste System and its Implications. Mark Sainsbury.* Louis Dumont and Basia Gulati, trans. Chicago: University of Chicago Press.

Durkheim, Emile. 1995[1912]. *The Elementary Forms of Religious Life.* Karen E. Fields, trans. New York: Free Press.

Ewing, Katherine Pratt. 1997. *Arguing Sainthood: Modernity, Psychoanalysis, and Islam.* Durham: Duke University Press.

Feher, Michael, with Ramona Nadoff and Nadia Tazi, eds. 1989. *Fragments for a History of the Human Body.* Vols. 1–3. New York: Zone Books.

Fernea, Elizabeth W., and Robert A. Fernea. 1997. Symbolizing Roles: Behind the Veil. In *Conformity and Conflict: Readings in Cultural Anthropology.* 9th edition. James P. Spradley and David W. McCurdy, eds., pp. 235–242. New York: Longman.

Findly, Ellison Banks. 1985. Gargi at the King's Court: Women and Philosophical Innovation in Ancient India. In *Women, Religion, and Social Change.* Yvonne Yazbeck Haddad and Ellison Banks Findly, eds., pp. 37–58. Albany: State University of New York Press.

Foucault, Michel. 1980[1978]. *The History of Sexuality.* Vol. 1, *An Introduction.* Robert Hurley, trans. New York: Random House.

———. 1985. *The History of Sexuality.* Vol. 2, *The Use of Pleasure.* Robert Hurley, trans. New York: Random House.

Gellner, David N. 2002. *The Anthropology of Buddhism and Hinduism: Weberian Themes.* New Delhi: Oxford University Press.

Ghurye, G.S. 1995[1953]. *Indian Sadhus.* Bombay: Popular Prakashan.

Gold, Ann Grodzins. 1984. *Fruitful Journeys: The Ways of Rajasthani Pilgrims.* Berkeley: University of California Press.

———. 1992. *A Carnival of Parting: The Tales of King Gopi Chand and King Bharthari as Sung and Told by Madhu Natisar Nath of Ghatiyali, Rajasthan, India.* Berkeley: University of California Press.

Gross, Robert Lewis. 1993. *The Sadhus of India: A Study of Hindu Asceticism.* Jaipur and New Delhi: Rawat Publications.

Gupta, Akhil, and James Ferguson. 1992. Beyond "Culture": Space, Identity, and the Politics of Difference. *Cultural Anthropology* 7(1):6–23.

Gutschow, Kim. 2003. *Being A Buddhist Nun: The Struggle for Enlightenment in the Himalayas.* Cambridge, MA: Harvard University Press.

Gyatso, Janet. 1987. Down with the Demoness: Reflections on a Feminine Ground in Tibet. In *Feminine Ground: Essays on Women and Tibet.* Janice Willis, ed., pp. 33–51. Ithaca, NY: Snow Lion.

Hansen, Thomas Blom. 1999. *The Saffron Wave: Democracy and Hindu Nationalism in Modern India*. Princeton: Princeton University Press.

Hausner, Sondra L. 2002. Hindu Renouncers and the Question of Caste. Paper presented at the Annual Meeting of the American Anthropological Association. New Orleans, LA, November.

———. 2007. *Wandering in Place: The Social World of Hindu Renunciation*. Bloomington: Indiana University Press (in press).

Heesterman, J.C. 1964. Brahman, Ritual, and Renouncer. *Wiener Zeitschrift fur die Kunde Sud-und Ostasiens* 8:1–31.

———. 1993. *The Broken World of Sacrifice*. Chicago: University of Chicago Press.

Holmes, Kenneth, and Katia Holmes, trans. 1995. *Gems of Dharma, Jewels of Freedom: The Classic Handbook of Buddhism by Je Gampopa*. Forres, Scotland: Altea Publishing.

Inden, Ronald B., and Ralph Nicholas. 1977. *Kinship in Bengali Culture*. Chicago: University of Chicago Press.

Jackson, Carl T. 1995. *Vedanta for the West: The Ramakrishna Movement in the United States*. Bloomington: Indiana University Press.

Jaffrelot, Christophe. 1996. *The Hindu Nationalist Movement in India*. New York: Columbia University Press.

Joshi, Sushma. 2005. Buddhist Monastery at Peace Village. <http://www.kantipuronline.com/artha.php?&nid = 49076>

Kaplan, Caren, and Inderpal Grewal. 2002. Transnational Practices and Interdisciplinary Feminist Scholarship: Refiguring Women's and Gender Studies. In *Women's Studies on Its Own*. Robyn Wiegman, ed., pp. 66–81. Durham and London: Duke University Press.

Khandelwal, Meena. 1997. Ungendered Atma, Masculine Virility and Feminine Compassion: Ambiguities in Renunciant Discourses on Gender. *Contributions to Indian Sociology* 31(1):79–107.

———. 2001. Sexual Fluids, Emotions, Morality: Notes on the Gendering of Brahmacharya. In *Celibacy, Culture, and Society: The Anthropology of Sexual Abstinence*. Elisa J. Sobo and Sandra Bell, eds., pp. 157–179. Madison: University of Wisconsin Press.

———. 2004. *Women in Ochre Robes: Gendering Hindu Renunciation*. Albany: State University of New York Press.

King, Richard. 1999. *Orientalism and Religion*. New York: Routledge.

King, Ursula. 1984. The Effect of Social Change on Religious Self-Understanding: Women Ascetics in Modern Hinduism. In *Changing South Asia*. K. Ballhatchet and D. Taylor, eds., pp. 69–83. London: School of Oriental and African Studies.

Klein, Anne. 1995. *Meeting the Great Bliss Queen: Buddhists, Feminists, and the Art of Self*. Boston: Beacon Press.

Knight, Lisa. 2005. Negotiated Identities, Engendered Lives: Baul Women in West Bengal and Bangladesh. PhD dissertation, Department of Anthropology, Syracuse University.

Lacqueur, Thomas. 1990. *Making Sex: Body and Gender from the Greeks to Freud*. Cambridge, MA: Harvard University Press.

Lamb, Sarah. 2001. *White Saris and Sweet Mangoes: Aging, Gender, and Body in North India*. Berkeley: University of California Press.

Leslie, Julia, ed. 1991. *Roles and Rituals for Hindu Women*. Rutherford, NJ: Fairleigh Dickinson University Press.

Mackenzie, Vicki. 1998. *Cave in the Snow: A Western Woman's Quest for Enlightenment*. London: Bloomsbury.

McDaniel, June. 1988. *Madness of the Saints: Ecstatic Religion in Bengal*. Chicago: University of Chicago Press.

McKean, Lise. 1996. *Divine Enterprise: Gurus and the Hindu Nationalist Movement*. Chicago: University of Chicago Press.

Moran, Peter. 2004. *Buddhism Observed: Travelers, Exiles and Tibetan Dharma in Kathmandu*. New York: RoutledgeCurzon.

Narayan, Kirin. 1989. *Storytellers, Saints, and Scoundrels: Folk Narrative in Hindu Religious Teaching*. Philadelphia: University of Pennsylvania Press.

———. 1993. Refractions of the Field at Home: American Representations of Hindu Holy Men in the Nineteenth and Twentieth Centuries. *Cultural Anthropology* 8(4):476–509.

Narayan, Uma. 1997. *Dislocating Cultures: Identities, Traditions, and Third-World Feminism*. New York: Routledge.

Neill, Roderick. 1970. Sadhus and Hippies. *Quest* 65(April–June):20–27.

Obeysekere, Gannanath. 1981. *Medusa's Hair: An Essay on Personal Symbols and Religious Experience*. Chicago: University of Chicago Press.

O'Flaherty, Wendy Doniger. 1980. *Women, Androgynes, and Other Mythical Beasts*. Chicago: University of Chicago Press.

Olivelle, Patrick. 1992. *Samnyasa Upanishads: Hindu Scriptures on Asceticism and Renunciation*. New York: Oxford University Press.

Ostor, Akos, Lina Fruzzetti, and Steve Barnett, eds. 1982. *Concepts of Person: Kinship, Caste, and Marriage in India*. Cambridge, MA: Harvard University Press.

Pechilis, Karen. 2004. Introduction: Hindu Female Gurus in Historical and Philosophical Context. In *The Graceful Guru: Hindu Female Gurus in India and the United States*. Karen Pechilis, ed., pp. 3–49. New York: Oxford University Press.

Pinch, William R. 1996. *Peasants and Monks in British India*. Berkeley: University of California Press.

Pollock, Sheldon. 2003. *Literary Cultures in History: Reconstructions from South Asia*. Berkeley: University of California Press.

Raheja, Gloria G., and Ann Grodzins Gold. 1994. *Listen to the Heron's Words: Reimagining Gender and Kinship in North India*. Berkeley: University of California Press.

Ramanujan, A.K. 1982. On Woman Saints. In *The Divine Consort: Radha and the Goddesses of India*. John Stratton Hawley and Donna Marie Wulff, eds., pp. 316–324. Berkeley: Berkeley Religious Studies Series.

Ramaswamy, Vijaya. 1992. Rebels-Conformists? Women Saints in Medieval South India. *Anthropos* 87:133–146.

———. 1996. *Divinity and Deviance: Women in Virasaivism*. Delhi: Oxford University Press.

———. 1997. *Walking Naked.* Shimla: Indian Institute of Advanced Study.
Schuler, Sidney Ruth. 1987. *The Other Side of Polyandry: Property, Stratification, and Nonmarriage in the Nepal Himalaya.* Boulder, CO: Westview Press.
Shaw, Miranda. 1995. *Passionate Enlightenment: Women in Tantric Buddhism.* Princeton, NJ: Princeton University Press.
Simmer-Brown, Judith. 2001. *Dakini's Warm Breath: The Feminine Principle in Tibetan Buddhism.* Boston: Shambhala.
Sinclair-Brull, Wendy. 1997. *Female Ascetics: Hierarchy and Purity in an Indian Religious Movement.* Surrey: Curzon Press.
Sobo, Elisa J., and Sandra Bell. 2001. *Celibacy, Culture, and Society: The Anthropology of Sexual Abstinence.* Madison: University of Wisconsin Press.
Strauss, Sarah. 2005. *Positioning Yoga: Balancing Acts Across Cultures.* Oxford and New York: Berg.
Tatia, Nathmal, trans. 1994. *Tattvartha Sutra ("That Which Is") by Umasvati.* The Sacred Literature Series. San Francisco: Harper Collins Publishers.
Thapar, Romila. 1979. Renunciation: The Making of a Counter-Culture? In *Ancient Indian Social History: Some Interpretations.* Romila Thapar, ed., pp. 63–104. Delhi: Orient Longmans.
Trawick, Margaret. 1990. *Notes on Love in a Tamil Family.* Berkeley: University of California Press.
Tsomo, Karma Lekshe. 1999. *Buddhist Women Across Cultures.* Albany: State University of New York Press.
Vallely, Anne. 2002. *Guardians of the Transcendent: An Ethnography of a Jain Ascetic Community.* Toronto: University of Toronto Press.
van der Veer, Peter. 1988. *Gods on Earth: The Management of Religious Experience and Identity in a North Indian Pilgrimage Center.* London: Athlone Press.
———. 1989. The Power of Detachment: Disciplines of Body and Mind in the Ramanandi Order. *American Ethnologist* 16(3):458–470.
Visweswaran, Kamala. 1994. *Fictions of Feminist Ethnography.* Minneapolis: University of Minnesota Press.
Wadley, Susan. 1977. Women and the Hindu Tradition. In *Women in India.* Doranne Jacobson and Susan Wadley, eds., pp.113–139. New Delhi: Manohar.
White, David Gordon. 1996. *The Alchemical Body: Siddha Traditions in Medieval India.* Chicago: University of Chicago Press.
———. 2003. *Kiss of the Yogini: Tantric Sex in its South Asian Contexts.* Chicago: University of Chicago Press.
Willis, Janice D. 1985. Nuns and Benefactresses: The Role of Women in the Development of Buddhism. In *Women, Religion and Social Change.* Yvonne Yazbeck Haddad and Ellison Banks Findly, eds., pp. 59–85. Albany: State University of New York Press.
———. 1989. *Feminine Ground: Essays on Women and Tibet.* Ithaca, NY: Snow Lion.
Wilson, Liz. 1996. *Charming Cadavers: Horrific Figurations of the Feminine in Indian Buddhist Hagiographic Literature.* Chicago: University of Chicago Press.
Wolf, Eric R. 1982. *Europe and the People Without History.* Berkeley: University of California Press.

Wolf, Margery. 1992. *A Thrice Told Tale: Feminism, Postmodernism and Ethnographic Responsibility*. Stanford: Stanford University Press.

World Bank. 2005. *Gender and Social Exclusion Assessment*. Kathmandu: World Bank and Department for International Development (DfID).

Young, Katherine. 1987. Hinduism. In *Women in World Religions*. Arvind Sharma, ed., pp. 59–103. Albany: State University of New York Press.

Young, Serinity. 1994. Gendered Politics in Ancient Indian Asceticism. *Union Seminary Quarterly Review* 48(3–4):73–92.

Portraits of Singular Women

Chapter 2

Do Saints Need Sleep? Baiji's Renunciation as Service

Meena Khandelwal

Baiji is a *sannyasini* who was initiated into the Saraswati Dashnami order as a young woman.[1] She has a large ashram of her own in Haridwar, a smaller one higher in the Himalayas, and many lay devotees. Baiji was reputed to be a strict, rather orthodox guru, but also generous in her interactions with both lay persons and other sadhus. In person, she was soft-spoken and quick to smile; she wore ankle-length robes and covered her short hair only for warmth in the winter. On our first meeting, she not only agreed to be interviewed but also warmly offered me a room at Rishi Ashram so that I might see for myself what life there was like.[2]

While many ashrams provide little more than shelter and two basic meals a day for residents, Rishi Ashram felt more like a family home. All residents, including servants, Brahmin priests (*pandits*), disciples, and guests received three full meals as well as morning and afternoon tea from the kitchen, and even packed lunches when necessary. Baiji did not require attendance at religious functions or allegiance to her religious teachings. Yet, most visitors came to Rishi Ashram solely to be near their guru Baiji, either to seek her advice and blessings or simply "to serve." She was not only the ashram's spiritual leader but also and unambiguously its administrator. Not all gurus are as involved as Baiji in the mundane details of managing their ashrams. Baiji made all final decisions on matters of food, finances, medical treatment, ritual protocol, and interpretation of scripture. She was not only the center of all activity but also the final authority on both spiritual and administrative matters. The meanings and activities of Baiji's service to others, in the context of her bustling Haridwar ashram and circle of lay followers, are the focus of my attention here; her Himalayan ashram is a much quieter place.

The multitude of ongoing charitable activities also made the Rishi Ashram busier, louder, and more cluttered than most other ashrams. The ashram provided room, board, a small allowance, and in some cases a college education to five or six young Brahmins from poor villages. In return, they maintained the temple, performed ritual services for Baiji's disciples, and helped with chores around the ashram. The ashram also ran a small Ayurvedic clinic, offered sewing classes to neighborhood girls, and founded clinics and schools in remote mountain villages. These services were offered to the poor free of charge. The extent of Baiji's involvement in these various projects was evident in the clutter that filled every inch of her ashram. In the hall there were huge sacks of grains and vegetables, bottles of pickles, tins filled with sweet and salty edibles, bolts of cloth, and straw mats spread out on empty floor space and covered with seasonal greens that were dried and stored for later use. Along the walls were rows of sagging shelves filled with health tonics, multiple copies of the Bhagavad Gita (an important Hindu scripture that is part of the Mahabharata), and years' worth of accounting ledgers. Baiji's involvement in charitable projects required her participation in a wide range of worldly activities such as negotiating with merchants to obtain cheap cloth for the sewing school, planning meals that met the needs of frugality, taste, and nutrition, hiring teachers and doctors, and recording financial transactions.

In addition to the charitable projects, Baiji's circle of devotees also pulled her into a range of worldly activities such as marital matchmaking and visiting homes. For Baiji providing guidance and other forms of service was the purpose of her *sadhana* (spiritual practice) rather than its byproduct. Her followers were generally middle- and upper-class and upper caste people. Although Baiji herself spoke no English, most of her closest followers, including the women, were as comfortable with English as they were with Hindi or Punjabi. The husbands were educated, some abroad, and were generally successful in business. Though not all of Baiji's disciples fit this elite class profile, those who formed the core of her following did. They hosted Baiji when she visited Delhi, and it was their wealth that built Rishi Ashram and supported its various charitable projects.

Ritual activities were equally central to the daily functioning of Rishi Ashram, and ironically they too involved Baiji in worldly matters, as she had to purchase ritual objects, prepare special foods, and plan what should be given as donation (*dan*). She looked after the priests by arranging their daily food, providing education, and even disciplining them. They were, after all, young men and had to be instructed to clean their rooms, behave with proper decorum, and attend to their studies. Moreover, many rituals were sponsored by householder disciples for specific worldly purposes: to mark a death anniversary; to celebrate an engagement; to seek divine intervention in resolving some crisis in health or employment. Baiji's devotees saw her involvement in worldly activities as *seva* (service). She offered food, shelter, and knowledge to me as a form of *seva*, so that I might fulfill my academic goals and serve as a conduit for what she considered to be Hinduism's highest truths. Radha Giri described by Hausner (this volume) lacks Baiji's formal religious training and high-caste status, but both women share a concern for serving the community.

Although Baiji is an initiate of the elite Dashnami order, she consistently refused to define herself according to many of the conventional categories employed in the academic study of Hinduism. Indeed, her religiosity crossed several conceptual boundaries. For example, it was not that she considered the distinction between *bhakti* (devotion), karma (action or good works), and *jnana* (knowledge) to be meaningless but that she followed all three paths rather than any one of them. Also, Baiji was considered "orthodox" by many who knew her because of her initiation by a Dashnami *sannyasi* (Hindu renouncer), her knowledge of Sanskrit scriptures, and her meticulous attention to ritual, and yet she also promoted the social service ("*seva*") characteristic of modern renunciant discourse. Another distinction blurred in Baiji's religious life is that between *nirguna* worship of an abstract divinity without form and *saguna* worship of anthropomorphized deities.[3] Devotional worship is particularly associated with the Puranas, ancient texts that contain stories about the personalities, family lives, adventures, and ritual practices of gods and goddesses. Devotion to a personal deity is not generally associated with *sannyasa*; indeed, some would find it antithetical to *sannyasa*, even though many renouncers participate in devotional forms of worship. Rishi Ashram housed both a lavishly maintained temple for the worship of personal deities and a sacrificial altar for the performance of fire sacrifice (*havan*). My observations of religious life in Rishi Ashram supported Baiji's self-proclaimed eclecticism.

Her Story: From Arya Samaj to Devotional Worship

Baiji's religious life is hardly a simple continuation of her upbringing. While she gives importance to having grown up in a religious household and in the proximity of saints, she has made a rather profound philosophical break with her past. Baiji was raised in a staunch Arya Samaj family that performed Vedic ritual daily and rejected the worship of images, but she now performs Vedic sacrifice as well as *murti puja* (the worship of icons) with apparently equal fervor. Her ashram maintains a large temple housing icons of Shiva, Vishnu, the Goddess Durga, and their various incarnations. The brightly painted and ornamented deities are worshipped through song and prayer for an hour and a half each morning and evening. The deities are offered incense and flowers, fed whatever is cooked in the ashram, bathed, and put to sleep at night. In addition, special *pujas* to major deities are performed on astrologically appropriate days of the week.

The Arya Samaj, a revivalist sect founded in the late 1800s by the *sannyasi* Dayananda Saraswati, aims to rid contemporary Hinduism of its "corruptions" (worship of icons, untouchability, ignorance of sacred texts) by returning to Vedic culture. The Arya Samaj valorizes such brahmanic ideals as vegetarianism, Vedic ritual, and Sanskrit study. It also, however, criticizes brahmanic orthodoxy's definition of caste as a status of birth. Instead, it considers caste to be an achieved status in the sense that any person who upholds brahmanic ideals in their daily life is a Brahmin, regardless of birth status. According to Baiji, her father was a devout Brahmin (and his name indicated he was Brahmin by birth) who practiced Ayurvedic medicine as a vocation, and their household was run according to brahmanic ideals. She describes her paternal grandparents as yogis who meditated for long periods. During childhood, she and her siblings woke before dawn each morning to meditate, perform fire sacrifice, and recite the Gayatri Mantra; only upon completion of their religious duties were the children given breakfast or allowed to leave for school. These habits of religious discipline, she says, were so ingrained in her personality that when the turning point came in her life she instinctively went in this direction.

A central belief of the Arya Samaj is that religious and educational gender equality characterized ancient Hinduism. Consistent with this, the Arya Samaj has historically encouraged both secular and Sanskrit education for girls in an attempt to improve the status of women. Also, both male and female members of the order are allowed to take *sannyasa* provided they are initiated by a guru of their own gender. Because Baiji was raised in an Arya Samaj atmosphere, when she began to desire a spiritual teacher she naturally imagined her future guru as a woman ("a Yogi Mataji") rather than a man. So it came as a surprise when, at age 22, she dreamt that a man put his hand on her shoulder and promised to teach her. She had not told anyone of her deep longing for a guru. Her father was in the habit of asking about his children's dreams each morning and when she related hers, he replied, "If you have decided about the path you must follow, then you will certainly meet him . . . Who is that man?" She did not recognize the person in her dream but replied that, from his appearance—he was wearing a white shirt and waistcloth—it looked as if he had come from the region of celestial beings (*divyalok*). His face was imprinted on her memory.

Six months later, the young Baiji and her family met a swami (lord or master) while visiting Haridwar, and even though his voice sounded familiar she did not recognize him at first. When he offered to take her father to meet a saint who lived deep in the forest she insisted on tagging along with the men. She asked the forest-dwelling saint if women could become yogis, and his reply was unambiguous: "Never." Just as the absoluteness of his answer began to sink in, and with it a feeling of despair, says Baiji, the swamiji who had escorted them into the forest put a hand on her shoulder and said, "Get up my child; I will teach you," *exactly* as had happened in the dream six months earlier. It was then that she realized he was the man who would be her guru. The next morning, and much to her family's chagrin, Baiji stubbornly refused to leave for the bus station until Swamiji kept his promise. At that moment, though, he was seated in meditation and therefore unavailable, and, with the bus about to leave, a battle of wills between Baiji and her father was imminent. Just then, Swamiji broke his meditation and called her and her father into his cottage. There he instructed Baiji to meditate on a certain mantra and sent her home. From then on, she began to meditate and keep periods of silence while living at home.

According to Baiji, it was through the meditation she practiced at home that she began to move toward the path of knowledge (*jnanamarg*) by understanding Vedanta philosophy intuitively rather than through the study of texts.[4] While this may resonate with a Western notion of intuition as a female way of knowing, Baiji never suggested that she considered intuition to be feminine. Moreover, her comment is entirely consistent with a general renunciant emphasis on experience over intellectual knowledge. Swamiji had her keep a spiritual diary and, through it, monitored her progress. Once Baiji and her family went to Haridwar to meet Swamiji with the idea that she would stay on for a while, but he was not there when they arrived. So, instead Baiji stayed at the ashram of a local Swami Ramanand, whom she had met through Swamiji on a previous trip to Haridwar. It would be safer, she was told, than staying alone at her guru's ashram. During this stay with Swami Ramanand she met a visiting *avadhut* who lived in a Himalayan pilgrimage town and eventually became her second guru.[5] "Avadhutji" formally introduced her to the philosophy of Vedanta and instructed her to memorize several scriptures, which she did. Eventually, she went to Uttarkhand where Avadhutji lived and remained there for several years, practicing *sadhana* under his guidance, until poor health forced her to return to the plains.

When I first met Baiji she was almost seventy and had had many bouts of illness. Yet her smooth skin, gleaming white smile, bright eyes, and abundant energy made it difficult to think of her as elderly. Her short, cropped hair was more black than gray and her movements quick, strong, and confident. A sadhu's youthful appearance is taken by many to be a sign of spiritual power. Baiji spent most of her time at Rishi Ashram in the late 1980s and early 1990s and visited her smaller, older ashram in Uttarkhand twice a year; more recently she has begun to spend more time at her Himalayan ashram.[6] Her circle of over thirty disciples was primarily composed of male and female householders who lived with their families. She had one (male) monastic disciple who had devoted himself full time to spiritual discipline rather than taking up family life. Here my focus is on Baiji's relationship with her lay followers, and particularly her close relationship with women disciples as expressed in the context of the "domestic" activities of the ashram.

Learning to Participate in the Life of Rishi Ashram

An ashram usually resembles the homes of its lay supporters. That observation was on my mind the evening I arrived at Baiji's ashram for my first ten-day visit. I already knew something about Baiji's disciples, having met several of them during the Ramayana reading I had attended several months earlier.[7] The wealth and aesthetic of Baiji's disciples were embodied in the building. It was modern, yet elegant in design. The outside walkways were smooth, made of polished stone rather than rough cement, and the external walls were coated with a tasteful whitewash instead of the more popular pastel blues, pinks, and yellows. These were the most superficial markers of the economic class of Baiji's followers. As soon as I slipped through the ashram gate, the gatekeeper took my bag and led me to the rear entrance, the one for daily, informal use. I left my footwear among the clutter of sandals outside the door and stepped onto the cool, slightly sticky stone floor of the deserted eating area. The lack of natural light made the gray interior look even grayer. Within seconds, Baiji appeared, greeted me, and instructed the gatekeeper to show me to my room, informing me that the bell would sound for the evening meal around eight o'clock.

Upstairs, the long, unswept hallway lined with padlocked doors indicated that the ashram was relatively empty. The rooms were new and inviting, each furnished with a ceiling fan, a bed, a tiny bedside table, a straw mat for the floor, and a single naked lightbulb. I was delighted to discover the luxury of a private bathroom, inside of which was a tap for bath water, two plastic buckets (one for fetching hot water and the other for mixing it in with the cold), a tiny sink, and a Western-style toilet, the type, no doubt, that one would find in the homes of Baiji's urban disciples. I recalled the large and crudely constructed dining table, covered with a vinyl tablecloth, which dominated the main room outside the kitchen. Though dining tables are a standard piece of furniture in middle- and upper-class urban homes I had never seen one in an ashram. In ashrams (except some catering to foreigners) it is customary to sit cross-legged on the floor for meals, and those who are either socially elevated or too arthritic to sit on the floor might be seated on a low stool with their food on a higher stool or small table in front of them. The large, communal dining table surrounded by chairs was a sign of the influence of Baiji's urban devotees.

Meals also followed a schedule more typical of urban households than of Haridwar ashrams, which adhere to an early-to-bed and early-to-rise ethos. That first night, the dinner bell rang at nine o'clock rather than eight. As it turned out, such late dinners were usual for Rishi Ashram. Since the priests were always served first, and guests second, the cooks and other workers, including Baiji herself, frequently had dinner as late as ten o'clock. I came downstairs at the sound of the bell to find the priests in the middle of their meal, seated on the floor in two rows. I was instructed to sit at the table, with Mehraji (an elderly and highly educated male resident of the ashram) and a visiting female disciple from Delhi. Baiji was not eating because of an upset stomach but sat at the table anyway to give us company and to supervise our meal. After eating the four of us remained at the table chatting while the cooks served themselves and sat on the floor mats, long since vacated by the priests, to eat. It was past eleven when we finally rose from the table, each of us washing our own utensils and then retiring to our rooms. Baiji had informed me that the morning bell would sound at half past three to rouse the priests so they could fire up the wood stove to heat water and bathe before the six o'clock prayer session, but that I could come down whenever I was ready. I rarely attended the morning service (few people did) but joined the evening prayer session most days and participated in a variety of ashram activities: performing *puja* along with everyone else, attending recitations of scripture, engaging in religious discussions.

While all this activity in the temple was important, the transformation of my status from a "guest" and outsider to an integrated, if temporary, member of the ashram resulted more from my involvement in ordinary activities than in specifically religious ones. I trace the development of this new role to my fifth day at Rishi Ashram. I had come downstairs some time before dinner to find Baiji and several priests engaged in the task of hand-stitching thin mattress pads from dirty, old burlap bags. A new gardener was recently employed by Baiji after his wife came begging for a job; Baiji came to know that the couple and their infant were sleeping on the wooden beds provided to them without anything underneath for warmth and padding, and though it was only September the nights had been cold and damp. She turned to me and smiled, "Burlap is supposed to be very warm. When somebody sends a donation here nothing is wasted." The burlap bags had arrived at the ashram containing gifts of grains and vegetables. Baiji kept the sacks, sure that she would eventually find some use for them. And rather than simply handing the sacks over to the couple to fashion some sort of mattress for themselves she saw to it that they were made neatly and durably.

Until that evening my offers to help in the kitchen had been rebuffed, and, being unsure of orthodox rules governing food handling and preparation, I did not insist. Rules of purity and pollution were closely observed at Rishi Ashram, and, having already been reprimanded for ritual mistakes more than once, I wished to avoid further embarrassment. Stitching burlap, however, did not seem to be a task requiring either purity or skill. After much insistence, Baiji finally allowed me to help, remarking, "What will Mehraji think that I am making you do this dirty work?" Because of my status as an educated middle-class person and an American ("but from a good Hindu family"), I was placed in the category of one who did no manual labor. My participation in the literally if not ritually dirty task of stitching dusty old burlap bags marked a turning point in my status and my experience of Rishi Ashram. The next evening I was permitted to help one disciple iron the doll-like garments for the temple deities, whose clothes were changed twice a month. Since there were 16 *murtis* in the ashram's temple and each outfit consisted of several articles of clothing, it was no small task. Gradually, as I learned the proper way of doing things and as others felt more comfortable requesting my help, I became increasingly involved in a wide array of chores, even those of food preparation.

One afternoon I emerged from my room at teatime to find several people gathered in the kitchen. There was a lot of work to be done, they said, so I gulped my tea, performed a purificatory rite by rinsing my mouth with water, and joined the others in the kitchen. The two cooks were called Lakshmi Behn and Sita Behn, "Behn" being a term for "sister." Lakshmi Behn had just returned from vacation and was thus new to me. She would only allow me to help after one of Baiji's disciples convinced her that I was in fact "one who worked" (*kamkarnewali*); even then she decided that it was fine for me to chop carrots but unacceptable for me to do the dirty work of cleaning mud-encrusted potatoes. By the next week, however, she was more than willing to hand me a large bowl of muddy potatoes for scrubbing.

The amount of time and attention devoted to food at Rishi Ashram seemed equal to that in any Indian home. Baiji was praised for making delicious dishes from what is considered waste vegetation, such as the leaves of zucchini plants or banana peels. "Baiji lived in the forest in Uttarkhand doing *sadhana*," explained one resident. "Then they had to subsist on whatever plants were available, so she knows how to make all kinds of tasty things out of nothing." Frugality and efficient use of resources were paramount values in Baiji's ashram. Baiji accumulated foodstuffs for thoughtful—never haphazard—distribution among the 20–30 people living at Rishi Ashram at any given time. In deciding what and how much to serve for meals, it seemed, she carefully planned what would be needed for an upcoming *puja* or feast and saved portions accordingly. Nothing was ever wasted. Whenever I returned from town with a box of sweets, it would inevitably disappear into one of the food cupboards (screened to allow air circulation but keep out rodents) and reemerge a few days later when needed for a special occasion. It was part of Baiji's duty as the head of the ashram to make sure there would be food for the next guest and *prasad* (blessed food) for the next *puja*. It was her duty to accumulate food and other material things for appropriate distribution. Because of careful management and Baiji's interest in cooking, food was both abundant and delicious, and, for this reason, her ritual feasts attracted large crowds of holy people.

The activities of food preparation provided the context for much of my research at Rishi Ashram as I learned much in conversations carried out over the rhythmic sound of dull knives hitting wood. Many of the devotees took an interest in my research. Most considered it entirely appropriate that I should wish to write about their guru, though a few were more suspicious of my motives and sympathies. The disciples who spent extended periods of time at the ashram often participated in both the formal interviews and informal discussions I had with Baiji. I had been at the ashram over one week when Baiji granted the first tape-recorded interview. Three other residents remained for the entire two hours, seated with me on the floor; for them this was a rare opportunity to hear a spiritual discourse from their guru. Indeed, devotional accounts of particular saints often take the format of question-answer sessions with the guru. During the first few weeks I asked most of the questions, but they also participated by commenting, translating unfamiliar words for me, and occasionally posing their own questions. My open-ended queries about *sadhana* or philosophy often led Baiji to offer personal anecdotes or reflections, although it is generally rude to ask a renouncer personal questions about their life before *sannyasa*.

It was after our first interview that Baiji began to refer to my previous life as a sadhu and how the habits (*samskaras*) of my past lives led me to choose this research topic. Baiji also began to remark often on how she could hardly tell I was a foreigner. Comments about my "Indianness" or my past life as a yogi were usually made in the context of an observation of how much work I did around the ashram or how well I served food. It was not the *amount* of work, since others did more than I, but the reversal it implied that inspired these comments. To them, I was a Westerner from a wealthy country, a scholar, and a respectable girl from a high-caste family, so the humility of my labor was perhaps more significant. Once, when the kitchen was short of help and I alone served the lunch Sita Behn had cooked for a visiting family of seven, Baiji triumphantly informed the guests at the end of the meal of my true identity as a foreigner. The grandfather expressed his happy astonishment that I was from abroad and still did so much *seva*. Occasionally, Baiji explained, it happens that the experiences from some past life manifest themselves in a present life. Thus, she continued, I must have been a yogi in a previous life and was born into a family in America in order to experience (*bhog*) some karma or another, but that the influence of those past lives led me back to India to live in ashrams and listen to the wise words of Indian sadhus. The assumption was that Americans would be reluctant to do physical work and that my willingness made me not only more virtuous but more Indian. Because mundane work was so valued at Rishi Ashram and was the locus of social activity involving almost everyone (Baiji, servants, disciples, priests), it offered an obvious opportunity to get involved in the ashram's social life.

Regardless of our differing interpretations of my participation in ashram life, that Baiji could forget I was American clearly delighted her. Yet, in crucial ways Baiji treated me differently from her women disciples. She assigned me to my own room upstairs rather than a communal one downstairs, ostensibly because I was a student in need of quiet. Also, she gave me a lot of freedom, realizing that it was necessary for me to move around a lot, often alone, in order to complete the research for which I had come. Baiji never expressed any disapproval at my going off alone to run errands in town or to meet other *sannyasinis*. I once asked if she minded my going places alone. "No," she said, "because you are in the habit and because you know so many people here and go only to familiar places." One day when I returned from town one of the cooks asked if my "man" would come from America to pick me up when I finished my work here. Baiji and I looked at each other and burst out laughing. "But she is a man!" exclaimed Baiji. Indeed, I played the role of a male in some ways. For example, when going to town I often ran errands for the women, who never ventured alone onto public transport, or was appointed escort to female visitors who wanted an excursion. Yet Baiji's comment was intended as a joke, for she did not really view me, or *sannyasinis* who wandered fearless and alone for that matter, as masculine.

That manual work rather than spiritual knowledge or devotional fervor inspired comments about my previous life as a sadhu is consistent with the value Baiji places on social service and physical labor. In this respect, despite her orthodox reputation, her lifestyle and philosophy are deeply influenced by a morality that is modern and utilitarian and that emerged in the context of British colonialism. In Haridwar's more orthodox ashrams renouncers are expected to devote themselves to spiritual pursuits, and work is often done by paid employees or junior monastic residents. At Rishi Ashram, everyone washed their own utensils after a meal, while hired cooks washed only the pots and pans. Although devotees or cooks (never priests) usually grabbed Baiji's used dishes before she could clean them, the disciples frequently remarked that Baiji washed her own clothes daily.

By the end of my initial ten-day visit I felt like a member of the ashram. That I was considered neither a stranger nor a nuisance by the people at Rishi Ashram made research there more fun and more fruitful. After six weeks back home in the United States my plan was to return to stay with Baiji for several months. I left with the impression that Rishi Ashram was as much like a middle-class home as an ashram. And Baiji was like the lady of the house, constantly supervising servants, preparing and distributing food, welcoming guests, and treating illnesses. When the time for my departure arrived, Baiji graciously invited me to return whenever and for however long I wanted. "There will always be room for you here," she said. "Consider this ashram to be your own home."

Sonam: Fierce Devotion

I soon returned to Rishi Ashram, arriving early one cold November morning by scooter from the train station. At breakfast, I found myself seated beside Sonam, a young Indian woman about my age. We quickly became acquainted. She lived in Bulgaria where her husband was on a three-year government posting. There she fell ill and spent four months in the hospital, and, even after performing all sorts of tests, the medical doctors were unable to find anything wrong. Meanwhile, she had become so weak that she could not even lift her four-year-old daughter Shivapriya. Baiji suggested that Sonam come to India with her daughter and stay in the ashram for treatment. Having diagnosed the illness as related to the spleen, Baiji prescribed a 40-day treatment that included drinking the juice of tender wheat grass, which Baiji was growing in pots especially for Sonam, and a glass of fresh cow urine (*gaumutra*) every morning. "How did you come to know Baiji?" I asked, after we had exchanged introductory information.

"My mother-in-law was a devout follower of Baiji's, and my husband and I also consider her to be our guru."

"Oh, so you only came to know Baiji after marriage?"

"Well, I myself am from Uttarkhand, where Baiji has an ashram, and I used to go with my parents to see Baiji on birthdays; we would take *prasad* and offer it to Bhagwan. We did not consider Baiji a guru but would visit her as one visits any saint or *mahatma*, to hear good words or just to be in their presence." It was almost a week later when Sonam explained to me how Baiji was actually responsible for her marriage. "Whenever anyone used to visit Baiji's ashram in Uttarkhand, she would have my father show them around town. A boy Bharat had come with his mother to visit Baiji. Though I did not know it at the time, his parents were looking for a girl for him; they wanted a strict vegetarian girl. I was twenty, qualified to teach and had not even thought of marriage. My father showed them around like so many others and then brought them home for tea. Within a week everything was fixed. Bharat's attitude was 'Whatever you say, Baiji . . . I only want your *prasad*; give me what you will.' So our marriage took place with Baiji's blessings."

Sonam smiled shyly and went on to explain that her mother-in-law, who passed away a couple of years before, was Baiji's actual sister.

"Oh, that would make Baiji your *masi* (mother's sister)!" I exclaimed.

"Yes," she said quickly, "but we don't think of the family relationship anymore after someone has taken *sannyasa*. Instead, we think of her as our guru."

Some Hindu householders continue their relationship with a relative who has taken *sannyasa*, although they cease to call that person by kin terms and the nature of the relationship changes. It is not unusual for a renouncer to take on a relative as a lay or even a monastic disciple. For example, one's sister or nephew may become a disciple. Usually, neither party will discuss the relationship in kin terms, although the ambiguity of the connection may be expressed in some contexts. Disciples, for instance, may take pride in their familial connection to a respected renouncer and seek to make the relationship known, though they would not do so in the presence of their guru. I once visited the cottage of an orthodox renouncer and was surprised to find his wife there. I realized she was his wife because of prior knowledge of the family, not because of her behavior toward him. She told me that she had come from Delhi for a month to cook and serve "Swamiji." Once, when she referred to "Swamiji" during a private conversation, I asked, "Which Swami?" "*My* Swami" she answered, to indicate her husband. Similarly, during a conversation with an elderly *sannyasini* whose husband had also taken *sannyasa*, Baiji referred to the woman's husband as "*your*

Swami." Regarding the fact that Sonam had married Baiji's nephew, Sonam's mother once remarked on the irony that for her coming to Baiji's ashram is like visiting her in-laws.

Sonam was very devoted to her guru and sought her advice on spiritual, health, and mundane matters. During her stay at Rishi Ashram, Sonam never expressed desires about what or how much she wanted to eat of any particular food. Before taking seconds of vegetables, for example, she would ask Baiji whether she should eat more. She consulted Baiji on everything from how she should discipline her child to what colors of clothing best suited her complexion. Once, Sonam obtained some cloth remnants from which she intended to sew some dresses for her daughter, but she wanted Baiji's advice on which cloth designs should be combined for aesthetic effect. It was days before Baiji had a chance to look over the cloth pieces, but Sonam waited patiently. Baiji would also discipline her daughter Shivapriya for sucking her thumb or for being obstinate, and Sonam always observed these interactions with appreciative amusement. Sonam generally avoided making any decisions on her own when Baiji was available for consultation. If a young married woman takes a guru she may find her devotion to the guru to be in conflict with her role as wife and mother. For Sonam, however, since her husband was equally devoted to Baiji, she could be an ideal disciple and an ideal wife simultaneously. Baiji taught her female disciples that their husband's wishes were first priority. She never initiated any woman as a lay disciple without the permission of her husband. To do otherwise would be to create disharmony within families and to invite criticism.

Sonam and I spent many hours together at Rishi Ashram preparing raw vegetables for cooking or drying. Baiji had us chop huge amounts of leafy greens and cabbage when they were plentiful, spread them out on newspaper or straw mats, and set them in the sun to dry. The dehydrated vegetables might then be sent up to Baiji's Uttarkhand ashram where greens are unavailable in the winter or stored away for feeding unexpected guests. My other activities included slicing fruit for *prasad* in the morning when Sonam was occupied, serving hot breads and replenishing vegetables during meal-times, helping with accounts (the tedious rather than responsible aspects of this job), and attending the evening prayer session. In addition, I spent a lot of time pestering Baiji for interviews and religious discourses. Immediately following the evening prayer session, the priests, disciples, and I would get up from where we were sitting to prostrate before Baiji who was usually perched cross-legged on her raised wooden seat on one side of the room. After this, we might leave the temple or remain seated there at her feet. Sometimes, such moments developed into an informal lecture or discussion.

Baiji as Guru

One morning Baiji promised to speak with me "after Mrs. Malia's interview." Shortly after my arrival at Rishi Ashram Baiji had begun jokingly to refer to all sorts of verbal interactions as "interviews." Mrs. Malia was a recently widowed middle-aged urbanite and longtime disciple who was visiting for a couple of weeks. She wished to discuss some personal matters with Baiji, and the two of them pulled their chairs out into the sun. Having returned an hour later to find them still deep in conversation, I sat in the courtyard with a newspaper, trying to remain unobtrusive and out of earshot but easily available should the opportunity for our interview arise. After half an hour, Baiji came to me and said that she had finished with Mrs. Malia but that the sun had made her drowsy. She insisted, however, that all she needed was something to munch on and summoned the gatekeeper to bring some white radish from the garden. We sat down at the dining table with my tape recorder and a plate of sliced, salted radish between us. "I slept only three hours in the night," she said. "Baiji, go and sleep for a while," I pleaded, to ease my guilt. "We can do the interview another time." "What's the point? I'll only sleep five or ten minutes . . . You must write in your thesis that *sannyasinis* also get sleepy." Baiji's spiritual powers, in the eyes of her disciples, were not diminished by displays of humanity such as illness, hunger, or sleepiness.

Baiji's spiritual power is one reason her disciples give for choosing her as their guru. Through meditation she is reputed to gain knowledge about everything from the treatment of illnesses to the whereabouts of deceased persons. I once inquired about an empty chair on one side of the room that was decorated with flower garlands. I had noticed the priests making offerings to the chair during evening prayers and Baiji herself bowing to it every day. She explained that when the Bhagavata Purana (a text containing stories of Krishna's childhood) is being read a chair should be available for the god Krishna in case he wants to come and listen, or when any recitation of scripture is going on there should be a chair for the god as well, "that is, if we really truly believe that these gods exist. If we take their existence as truth, we should provide a chair for them." Similarly, the priests rolled out four straw mats during the *puja* and decorated them with flowers. Baiji said these are spread out in case anyone from the heavenly region (*paramlok*) wants to come and join in. Thus, they are brought out before the *puja* as an invitation and are removed afterwards so as to say, "O.K. it's over, you can go now." With a chuckle, she waved her hand as if shooing away a pesky god who had overstayed his welcome.

Baiji is said to have cured one disciple of cancer and another of infertility, and she treats herself for various ailments. Once she was sick with a stomach ache for a long time and had a vision to eat hot chili pepper on a bread roll! She was afraid of the burning it would cause, but, in an act of faith, spread a thick layer of chili pepper on a bread roll and ate it. Miraculously, she felt no burning in her mouth, eyes, lungs, or digestive track. After three days she had a vision that she had eaten too much and should stop; then she felt the burning in her mouth and stomach and while defecating. Red chili, she explained, like onions and garlic, kills bugs. Anyway, the stomach pains disappeared. In addition, Baiji is attributed with the ability not only to communicate with deities but to influence them as well. The most dramatic proof of this is the frequently told story of how she saved a Himalayan town from floods by performing a *puja* to Mother Ganges and asking her to retreat. Baiji's mystical powers are also expressed in everyday activities, in the knowledge, for example, of how to make tasty food "out of nothing" or how to draw up architectural plans. When I asked a disciple why something was done or how Baiji had learned something, the most common answer was "I don't know; she must have seen it in meditation."

I suggest, however, that Baiji's value as a guru derived as much from activities of a less glamorous nature. Her *seva* included all the mundane work of running an ashram: supervising projects, hiring and firing employees, managing the accounts, and organizing meals. One day Lakshmi Behn, Baiji, and I spent hours chopping mustard greens from the garden to be dehydrated and stored for later use. All three of us were getting tired of the work, and Lakshmi Behn was beginning to think up other chores that needed doing, hoping perhaps to find relief from the task at hand. Suddenly, Baiji began to giggle and told us an amusing story about a king whose stingy and tyrannical nature almost drove his own children to the point of murdering him. She was aware that we might think her too exacting at times, but this did not fluster her. Baiji seemed to accept that the job of running an ashram necessitated strictness and that this would sometimes be seen as cruel. Scoldings or periods of hard work were often followed by expressions of consideration or love. One night, when I started nodding off at the dinner table after a day of hard work, Baiji offered to massage me; had I accepted, her massaging me would have been a dramatic reversal of normal hierarchical relations between sadhus and householders. Baiji's disciples constantly remarked on how much work their guru did. The difference of course was that the cooks were euphemistically called "*sevaks*" rather than "servants" but were basically domestic workers, while Baiji's *seva* was a matter of greater choice and contributed to her reputation as a genuine saint.

Baiji's *seva* was not limited to domestic tasks, for charitable projects also kept her busy. However, on a daily basis, she was so involved in housekeeping, food preparation, providing hospitality to guests, and other aspects of ashram management that she had little time for recorded interviews. Thus, most of my research took place in the snatches of conversation that broke the tedium of repetitive chores. Sometimes I would get ready in the morning and venture downstairs early, around seven, in hopes of getting a little time with Baiji before she got involved in other tasks of the day. One such morning I found Baiji sitting in meditation in the temple, so I busied myself by cutting the morning's *prasad* of apples and bananas for distribution. By the time Baiji emerged from the temple breakfast was ready. After breakfast a malfunction in the gas stove had to be repaired so that lunch could be cooked. After seeing that it was fixed, Baiji began gathering food to send as *prasad* to her Uttarkhand ashram because someone was traveling there the next day. In rummaging through the cupboards Baiji found three boxes of sweets that had been long forgotten. Tragically, they had begun to mold around the edges. When she suggested that we scrape off the mold, then cook them again in clarified butter, I tried to look enthusiastic and nodded affirmatively. It had to wait though because Sita Behn was alone in the kitchen that day and vegetables needed to be chopped for lunch. Once the vegetables were cleaned, chopped, and handed over to Sita Behn for cooking, there was still time left for an interview. We settled down in Baiji's room with the tape recorder, but just as she became animated in her description of the different heavenly worlds, it was time for lunch. Baiji promised to continue the interview later. After lunch we started cleaning the moldy sweets and before long Baiji began to feel sleepy and got up to lie down, saying "What's your name—Meena—I'm sorry." This was typical of my days at Rishi Ashram. While helping in mundane tasks was a way of getting to know people and participating in ashram life, I relished those "interviews" as the rare times when I had Baiji's undivided attention.

Baiji reflected on her life now and then. "I am glad you came but sorry that I have been unable to serve you (*seva karna*). . ." she said to me one day. "When we go to Uttarkhand I will have more time to spend with you because there won't be all these ashram matters to worry about." She had invited me to accompany her to visit her Uttarkhand ashram the next month. She had a trusted man there who looked after the ashram, she said, "otherwise, it would become like the situation here." This was the first time I heard her express any dissatisfaction, however mild, with Rishi Ashram, though she always spoke of the Uttarkhand ashram fondly as if it were the place she would prefer to be.

Ritual Purity

Baiji's attitude toward ritual pollution at Rishi Ashram seemed inconsistent. Earlier in the month, Sonam was in a state of ritual pollution because of her grandmother's death. Baiji had told her to avoid the temple and kitchen for 13 days and, of course, not to touch food or cooking utensils. That morning when raw sugar was needed in the kitchen, Baiji sent Sonam with me to fetch it. Sonam led me to the appropriate cupboard and instructed me to pick up the sugar and carry it back to the kitchen. The next day, I noted that Sonam, on Baiji's request, went to retrieve the raw sugar herself. When Sonam was asked to make some carrot-apple juice for Baiji later that same day, I offered to operate the electric juicer. "No, this I can do," she had said. "I just cannot enter the kitchen . . . Since Baiji is from among the saint-sadhu people, all of these things are not so important for *her*. God is in everyone." Baiji, it seemed, as a renouncer was not affected by ritual pollution while the priests and deities were. Later that day, however, when Baiji saw that I had been chopping bitter melons alone for a long time, she sent Sonam to help me. The vegetables were to be cooked and offered to the deities and then fed to the priests along with everyone else. I was utterly confused.

With regard to menstrual pollution, Baiji also seemed inconsistent. When I first went to stay with her she told me that, while menstruating, women usually stayed in their rooms and she had food sent to them. One month, Sonam was prohibited from entering the kitchen but allowed to help me cut raw vegetables in the dining area. The next month, however, she ate meals in her room from a separate set of utensils that she washed in her bathroom. And, she was asked to keep her distance from the priests. One of these days, Baiji and Achariji (the head priest) were busy cutting bolts of cloth for waistcloths to be given as donations. I started helping Baiji cut the cloth, then I would courier it out to the front room where Sonam was hemming them with the sewing machine. Sonam could not enter Baiji's room because Achariji was there. In addition, she could not handle food, even raw vegetables, or sit at the kitchen table with us. When I asked her the reason for this sudden strictness she said, "I do what Baiji tells me. Before, Baiji had said that from drinking cow urine everything becomes pure." This month, however, things had changed. When I tried to question Baiji about the change in rules, the answer was vague. During menses, she said, the *vrittis* (inner dispositions) change and one's attention becomes focused on the body. Baiji's sentiment that menstruating women should not do mental or physical work had been echoed by other women renouncers I had met. In the village where Baiji used to live, the women were not allowed to cook food during their menses, yet they were expected to fetch water from the well and do other hard work. "That is not right," she said. "They should get three days of holiday."

Several months later when I accompanied Baiji to her Himalayan ashram I was surprised to discover that none of these rules of purity were followed. The cook entered the kitchen without bathing, and we householders were permitted to enter the temple before warm water was available for a morning bath! One morning, in response to my confusion, Baiji told a story about the sage Vashisht who stayed high in the Himalayas where few people live; it was a region mainly inhabited by sages and seers who were immersed in austerities. It is so cold there, she said, that one cannot even bathe in the icy water. As the story goes, the sage was happily eating food prepared by his menstruating wife Arundhati when he was called down to the plains by Ram's brother Lakshman. Heeding Lakshman's plea, the sage finished his meal and then, without even washing himself, started off on his descent. When they arrived at the plains of Haridwar, Lakshman watched aghast as his guru bathed in the Ganges, wore a clean waistcloth and a rosary, and applied ritual markings to his forehead—he had transformed himself into a Brahmin priest! When Lakshman finally built up the courage to point out this hypocrisy to his guru, the sage was amused. He explained that this place (Haridwar and the plains of North India) was the land of karma where one must pay attention to a system of rules, but that they had descended from the land of knowledge and austerities (the higher Himalayas) where there is no question of male or female, pure or polluted (Khandelwal 2004:128–129). This story was Baiji's way of explaining her erratic pattern of obeying and ignoring rules of purity as a matter of context, place, and appropriateness.

Behind the Scenes at a Ritual Feast

Baiji awoke one morning with flu-like symptoms. I made her some tea from basil leaves, argued for the curative powers of hot salt water, and then insisted that she rest after gargling rather than going into the temple (where the priests had requested her presence). When I joined a group of women chopping vegetables, one disciple Rajinder pointed to Baiji's room and said to me knowingly, "This [*seva*] is the real research, isn't it?" She stated that even though she had known Baiji since she was in the third grade and spent a lot of time with her, she still found her hard to understand. "Other *mahatmas* [great souls] are easier to understand," she said. "They may give only theoretical talks and philosophical advice, but Baiji gives practical advice. She is very 'self-dependent'. There is nothing she cannot do. She plays sitar skillfully and used to draw as well. She knows Ayurvedic medicine and construction work. Yet you can ask her about any Veda or Upanishad and she can teach you.[8] She learned a lot from her first guru but took *sannyasa* from Avadhutji and considers both to be her gurujis. Avadhutji was very strict, like a lion in the jungle. He ate once a day—fruit, sweets, or whatever—and would speak only about religion. Baiji performed very difficult austerities. Many people, especially *mahatmas*, opposed her because they did not like *sannyasa* being given to women. But he watched her do *sadhana* and could see that she had risen quite high. So he thought there could be no harm in giving *sannyasa* to such a person. He only made two disciples. Baiji did hard austerities, you know . . . She used to keep silent for six months at a time." Paradoxically, renunciation promises the possibility of transcending all social distinctions, including those of gender, while also institutionalizing male privilege in multiple ways.

Rajinder grew up in Himachal Pradesh where Baiji also lived with her family, and Baiji's sister was Rajinder's primary school teacher. She said she used to visit Baiji often when she was young and liked being around her, though Baiji would discourage her from coming, telling her to go home and study. Baiji's first ashram is there; it is small but has a primary school and dispensary. Rajinder chose not to marry. Instead, she taught in the school for six years and eventually became its principal. When Baiji was young, according to Rajinder, she used to play the sitar and slip into meditation. "Her family did not accept this side of her," she said, "except her father; he believed in it." I asked how Baiji was different from other *mahatmas*. "Other *mahatmas* will have their disciples worship and serve them, but Baiji serves her disciples. She does more work than anyone else here in this ashram. Other *mahatmas* will give orders, but Baiji never tells anyone what to do. Instead, she'll do it herself; then we will feel that we should also do something. Also, Baiji has practical knowledge. She can do anything and she herself supervised all the work of building the Uttarkhand ashram. She even drew up the [architectural] plans."

"How did she learn to do this?" I asked.

"I don't know," said Rajinder, "it must have come to her in meditation."

"Baiji is always ready to learn some new thing," added Sonam. "If a laborer comes she will want to learn from him—'Learn something,' she'll say."

"Also, for Baiji all are equal," Rajinder continued. "*Sannyasis* are supposed to see all people as equals and Baiji really does. We may want to give an important person better *prasad* and more attention, but Baiji treats all equally. If anyone comes, whether Brahmin or Chandal, she will want to give them some good *prasad*, and no matter how sleepy she is she will want to listen to their whole story. She will satisfy them not just physically but mentally as well."

"Chandal?" I asked, seeking fuller explanation.

"A Chandal is an untouchable," she said, switching from Hindi to broken English.

Sonam added, "A Chandal is one who does any wrong things."

"Yes," Rajinder corrected herself, "Chandal is not a quality of one's birth. In Baiji's first ashram there used to be feasts constantly, and she fed the *mahatmas* with such love. She even used to serve those *mahatmas* who opposed her in Uttarkhand when they were sick." Taking advantage of the conversational mood, I asked whether Baiji was ever married. The enthusiastic conversation ended abruptly.

"I don't know," Rajinder finally said to break the silence.

Sonam added a gentle reprimand. "There are some things one should never ask a *mahatma*. Do you remember that South Indian *sannyasini* who told you not to ask about her previous life before *sannyasa*?"

"Yes," I said contritely, reminded of my outsider status.

Rajinder skillfully changed the subject. "When I was young I used to get really disgusted with the latrine. Then one day I saw Baiji cleaning up the shit of a small child. At first I was revolted, but then I realized that if such a great *mahatma* was doing it then there must be nothing wrong in it." There was a chorus of praise for how hard Baiji works, especially considering that she was almost seventy and in poor health.

Baiji, meanwhile, was perched upright on her bed, eyes drooping and nose flowing like a river; instead of resting, however, she had begun a list of the necessary preparations for a feast (*bhandara*) to be held later in the week. Five hundred people were expected for the feast and no outside labor would be called in to help; it was going to be a lot of work for all of us. In response to my shock at the numbers, Baiji said that when one person hears about a feast at Rishi Ashram everyone wants to be invited (because they know it will be generous). After several minutes of scribbling and calculation she figured how many vats of each vegetable would be required and concluded that, if everything was to be ready in time, the pumpkin should be chopped and peeled tonight. Within minutes the word was out. We took our positions around the dining table, piled high with pumpkins, and hacked away at them late into the night.

The morning of the feast, I bathed and spent a couple of hours catching up on my field notes before venturing downstairs around eight o'clock. Everyone was working quickly but calmly. When I asked what needed to be done, Baiji had me work on a project completely unrelated to the day's activities. The morning after the feast she was leaving for Delhi where she would perform a seven-day recitation of the Bhagavata Purana for her disciples. She wanted me to copy the underlining from her old copy of the text into the new edition that she planned to carry with her. I took the colored markers and began highlighting the significant passages, until, a short while later, Baiji decided I should dice the carrots that I had left unfinished the night before. Completing that, I joined the group of ladies in the kitchen who were rolling balls of bread flour in the palms of their hands and then flattening them with a rolling pin for frying; communal tasks were always more fun than solitary ones. In the temple, the Bhagavata recitation was proceeding as we worked. "Baiji told me that sitting in the kitchen doing work is as good as attending the Bhagavata," said Rajinder cheerfully. "Half of my devotion is in the kitchen and half in the temple." Another devotee came and squatted beside me, breaking off a little dough and squashing it between her palms. She looked at me and said, "The yogurt was not ready on time. Whatever goes wrong always comes out on Baiji. If the water supply stops it is Baiji's problem. If the food is not ready on time and the *mahatmas* must wait, it falls on Baiji's head."

We were given a light breakfast at eleven, as the first batch of *mahatmas* was arriving. As soon as we finished, Baiji called Sonam, a younger girl, and me outside to help distribute cloth and two rupees to each of the renouncers seated in rows in the inner courtyard. They were all men, well groomed and respectable-looking. Ours was a ritual participation because Baiji had us each hand the donation to only three or four sadhus before calling someone else to take our place. As I turned to go in, I heard a woman's voice loudly reciting a Sanskrit verse on the other side of the gate and learned that some women renouncers had also arrived. I decided to stay outside and help with the serving. There were 13 *sannyasinis* in the next batch (none in any other), and they sat together segregated from the men. All were elderly except one. When I began distributing food to them, many asked for especially soft breads because they had few or no teeth. One woman, seeing that I was having trouble finding enough of the soft breads that everyone wanted, informed me with a smile that she liked hers hard and chewy. The *sannyasinis* were dressed in orange saris, or petticoats with long tunics, and all had their heads covered. Baiji stood among the renouncers, in the hot mid-day sun, supervising the amazingly efficient serving of food. That the sadhus were seated in long rows of 20 or 30 people facilitated the feeding process. The *mahatmas* did not eat as much as everyone predicted they would, apparently because there were four other feasts in the area that day. It took less than an hour for the hundred or so people to sit, eat, and leave. As they departed en masse, the banana leaves were removed and the driveway quickly swept clean for the next batch. When all the sadhus had come and gone, there was still lots of food left, so Baiji summoned 15 laborers who were working nearby to come eat. They were fed as the *mahatmas* had been.

Back inside the ashram, the priests and a couple of the male servants sat down for their meal and, afterwards, each of them was given a length of cloth and two rupees. Baiji told us to go ahead and eat without her because she was going to the altar where fire sacrifice was nearing its conclusion. It was two o'clock in the afternoon and the ritual had begun around nine thirty in the morning. We all followed Baiji into the altar rather than heading toward the dining area. Two "Aunties" who cosponsored the ritual to mark the deaths of kin were seated near the sacrificial fire while Achariji officiated. At the end everyone offered flowers into the fire, then bowed to Baiji. It was over. Baiji glanced around at the group and said, "Shall we go?" Achariji remarked that a few words should be said about this past week. He addressed the group in formal Hindi, expressing his hope that the activities of the last seven days would bring success to those for whom they were done; then he asked Baiji to say something. "Well, usually at this moment there is no talk." She paused, looking exhausted and waiting, no doubt, for her followers to let her go. But they continued to watch her expectantly. Finally, she closed her eyes and started speaking in a monotone voice that she only used in ritual contexts. She spoke about death and how ritual actions such as the one performed here today most certainly help our loved ones in the afterlife. She concluded by stressing that if we wish to perform certain actions to ensure a safe passage into the heavenly worlds, it is far better to do them while we are still alive. With this, the audience was satisfied and slowly began to move toward the door.

It was four o'clock when we seated ourselves around the table for lunch, and I devoured the rice, pumpkin, potatoes, and fried bread. Baiji wanted me to eat some fruit cream, a costly food, but the thought of eating it made my stomach turn. Baiji dropped a spoonful of it on my plate, and once it is on the plate it must be eaten. Attributing my reluctance to being too full rather than to taste, she suggested that I send some of it to my husband as *prasad*. "Just concentrate very hard and it will really go to him," she instructed. I closed my eyes for a moment and prayed that cream might disappear from my plate. I opened them again, blinked, looked at her and said "I tried, Baiji, but it is still here." Everyone exploded into laughter and Baiji explained, "It has to go from the *stomach*." As I prepared to down the little pile of fruit swimming in clotted cream, Baiji began to narrate a story about having to attend some *event* at an ashram while she was in Delhi. "The people at home [the disciples with whom she was staying] told me not to eat too much because I wasn't well. I went to the ashram and ate all the things I wasn't supposed to eat and felt fine. Then I came home and they asked if I had eaten anything, to which I responded 'No.' Then one of *them* suddenly threw up and all those things *I* had eaten—okra, etc.—came out. They found out what all I had eaten because it was all right there on the floor! So, it really is true that you can send the *prasad* to your husband."

"One time I went to a *path* (recitation) with another lady," Baiji continued. "Usually, this lady can hardly even eat two *rotis* (unleavened bread) but that day she ate 13! All the flour in the kitchen was finished and she went home feeling hungry. The hosts were dismayed because they felt they were sending their guest home hungry. I myself had only one *roti* that day and was feeling full just watching that woman eat. We went to the home where we were staying and they had warmed milk with clarified butter for me.

I couldn't drink it because I wasn't feeling well, so my companion drank it. Then, still feeling hungry, she went into the kitchen and ate the servants' big fat *rotis* as well!" Vimla Auntie turned to me and said, "These are facts that we have all experienced with Baiji."

Baiji Reflects on her Gurudom

One evening a few weeks later, after I had been pestering Baiji for an interview all day, she told me to come into her room with her and she would talk to me while she worked. She and Achariji were busy cutting bolts of cloth for waistcloths and preparing other things to be given in donation. Baiji was to carry all these things to Uttarkhand with her. Achariji started to light a candle to use for sealing a plastic bag. Baiji warned him not to use a candle by reciting a verse.

> *Agni, jal, sadhu aur raja*
> *In charon me bharosa nahin rakh sakte hen.*
> [Fire, water, sadhu and king
> In these four we cannot trust.]

As Achariji laughed, Baiji insisted, "It's true. With a flame, if a little wind comes along then it is all over. And a sadhu? Well, they might just get up and leave at any moment."

The next day we started on what turned out to be a two-day project: chopping vegetables and measuring spices for making Baiji's much-loved cauliflower and radish pickle. On the second day, as Baiji and I sat together to finish chopping the last heads of cauliflower, she asked me, "You'll write this in your thesis? Sadhus do what sort of work? Householder's work." We laughed. The moment seemed right to pose a question that had been on my mind since she recited the verse about how one cannot assume that a sadhu will stick around or stick to a plan. "Baiji, have you ever felt like leaving this ashram and running off to Uttarkhand?" She always spoke of the place so fondly and often said that there she had no headaches or work. "Yes," she responded quietly while glancing at her surroundings wistfully. "I used to think of that a lot, but now I try to calm my mind and tell myself that if it is God's wish, then . . ."

That evening a Brahmin priest who was visiting from Punjab joined the prayer session. Krishna used to be one of the young priests of Rishi Ashram but now worked as a priest elsewhere. After the ceremonies he and I both sat at Baiji's feet. He first reported on what he had done that day and his plan to return home the next day. Soon, it became clear that he wanted Baiji's advice. Someone had suggested that he get a Bachelor's degree in education and he was seriously considering it. Baiji encouraged him to go for the degree. She told me earlier that she thought very highly of him—that he was very knowledgeable. By then Achariji had also joined us. The priest mentioned to Baiji that Achariji advised him not to go for the degree and felt that this religious field is the best one. Baiji, however, was more pragmatic. "Now in Delhi," she said, "there are about four hundred ritual officiants (*pujaris*) who are not of Brahmin birth, but they are performing *pujas* because they are educated with PhDs, etc. and because educated Brahmin priests are not available. And they are demanding a lot of money, too." She offered the guest further advice about performing religious services for householders.

"You must look at your own experience," she continued. "The hardest thing is not to get stuck in the egotism (*ahamkar*) of I am, I am doing, etc. A person may give 5,000 rupees one time and nothing the next, but you must remember that it is the same person. Then he may give 500 rupees the third time. It should make no difference to you. Another person may give nothing, but at least he's taken out the time to *come*. And you must not wonder why someone who gave money the last time is not giving any this time. After all, it was not *you* who took that first 5,000 rupees . . . The money was not for *you*. He gives it for himself. Whatever is given is not given to you but to God, so if you take it for yourself then you are stealing. You must not accumulate it or it becomes filth. The one who does not see any difference between those who give money and those who don't, the one who is unaffected by what people give, whether praise or criticism, that person is the real saint. It is very difficult to save yourself from egotism. If you can, then your life will be successful; otherwise you will go on being reborn. *Paramatma* (Ultimate Reality) is simple, but this worldly business (*vyavahar*) is not. It is very difficult to be in this worldly business and not be touched by it, but there are still some great saints around. There was one very great saint called Narayana who lived in Uttarkhand. He used to be a judge before renouncing worldly life. He only took alms in a small bowl from one house once every three days, and in between he did not even drink water. He spoke little and very quietly. If someone offered him money he neither looked at it nor touched it. He was very beautiful, with long black, matted locks."

Baiji spoke from experience, for she emphasized the difficulty in being involved in the world without being affected by it. She was certainly familiar with the dangers and pitfalls of the path of *seva*, which demanded that she live in the midst of householders and worldly concerns. At times Baiji would reflect on the pitfalls and temptations faced by a renouncer who devotes herself to serving the society of householders: the temptations to think of the donations as personal possessions rather than things held in trust (to be redistributed to the needy), the desire to run away from all the hassles for a quieter life, and, most of all, the ego produced by reverence. Yet Baiji seemed confident that she had chosen the right path, even if it was, in her estimation, more difficult than the path of contemplation and ascetic isolation.

Sannyasa as Service

If there is any single quality of Baiji's *sadhana* that colors most of her activities and teachings, it is an emphasis on action (karma), both ritual and charitable. Originally karma referred to the proper fulfillment of the ritual duties of a householder, so it represented a path renouncers had rejected. The Bhagavad Gita, which is part of the great epic Mahabharata and is one of the most influential texts in India today, assumes an active way of life and speaks in terms of yoga (application of self-discipline) rather than *moksha* (enlightenment) (Embree 1988:278). However, the Bhagavad Gita also addresses the problem of how one can attain liberation while continuing to act in the world and, thus, incorporates elements of renunciation. Lord Krishna advises Arjuna that one should perform an action without any concern for or interest in the results. Action that is free from attachment and self-interest is thus better than inaction. This ideal of acting with detachment legitimizes the work of actively engaged renouncers. Baiji had committed the Bhagavad Gita to memory, referred to it often, and held it up as a distillation of the Upanishads.

While the notion of renouncing the fruits of one's actions is found in the Bhagavad Gita, the interpretation of karma as charitable social service is modern. The valorization of selfless action in the Bhagavad Gita has been widely interpreted by modern renouncers as a call to work for the uplift of India's poor. Ursula King has argued that modern Hinduism, shaped in part by increased interaction with Europe, has reinterpreted the Bhagavad Gita so as to give individual initiative and action a new legitimacy. The notion of karma (literally action or good works), which in the Vedas referred specifically to ritual activity, later came to mean human action in a much wider sense. One implication of this change was a relatively new evaluation of physical work and social service, partly due to the influence of Christian missionaries who explicitly equated karma with service rather than ritual activity. King has noted that the Hindu social ethic of *karmayoga* (the yoga of good works), even if introduced by Christianity, can be grounded either in the absolute monism of *sannyasa* or in the belief in a personal god (King 1980:42–48). Thus, in its insistence on breaking down the boundaries between one person and another, Advaita Vedanta provides a philosophical justification for spontaneous service motivated by compassion for the pain of others. Alternatively, all good works can be interpreted as an offering to a personal god. Dazey (1990:311–313) has discussed the way in which new monastic institutions ("committed organizations") have incorporated a broad range of social services including medical aid, famine relief, literacy campaigns, and village development, through complementarity and hierarchy of spiritual disciplines. In accordance with this, Baiji, like all renouncers I met, held that one cannot reach liberation through service alone; good works must be combined with austerities (*tapasya*) and knowledge (*jnana*). It was also clear that she did not view the classical and modern meanings of karma (ritual activity and social service) as mutually exclusive since both are an important part of religious life at Rishi Ashram.

This modern reinterpretation of *sannyasa* emerged during a period of social change. During the eighteenth and nineteenth centuries, British colonial officials and missionaries targeted sadhus, along with ritualism, untouchability, child marriage, and the plight of widows as evidence that India was mired in irrational spiritualism and regressive social values and in need of Britain's "help"; the sadhu was thus perceived as a drain on India's meager resources—even as a reason for its poverty (Narayan 1993). Partly as a response to this onslaught of criticism both educated Indians and ascetics themselves began to envision a new socially "relevant" form of Hindu renunciation. Swami Vivekananda was a key figure in this reinterpretation of *sannyasa*. He represented Hinduism at the 1893 World's Parliament of Religions in Chicago, after which he toured the United States for several years, established the Vedanta Society, and initiated Western disciples. Vivekananda articulated for Western audiences a vision of Hinduism that was spiritual, tolerant, and universal—and that supported the Orientalist opposition between Spiritual East and Materialist West. More importantly, and in contrast to the teachings of his guru Ramakrishna, Vivekananda believed that renouncers could help in a national effort to uplift the poor rural masses.[9] He redefined the ideal renouncer as committed to India's cultural, material, and spiritual renewal (Sinclair-Brull 1997:32) and as a participant in organized service to humanity (Beckerlegge 1998:188). Thus, although Baiji is reputed to be orthodox in her adherence to Vedic ritual and her knowledge of scripture, her focus on service is thoroughly modern.

During the last century, then, *sannyasa* has come to emphasize active involvement in society in the form of social service activities. As Kalyani Menon shows (this volume) this modern discourse of an active, engaged *sannyasa* is embraced by nationalist women to justify their political activities on behalf of the Hindu nation. This new emphasis on social service and reform has not come to *replace* the focus on disengagement from worldly life. On the contrary, many contemporary renouncers still consider activism to be a distraction from the goals of true renunciation and most think it impossible to attain liberation through social service alone. These critics are skeptical that social service can ever be truly disinterested and that a renouncer can maintain his or her detachment and lack of ego while being intimate with householders. While some renouncers may choose one or the other of these competing ideals, many attempt to balance them. The values of detachment and social involvement are held in tension, and individual renouncers hold varied opinions on the relative importance of each. Many try to resolve the tension by combining active involvement in social service with an *internal attitude* of detachment, though this is considered to be a very difficult balancing act to maintain.

Not only is Baiji's renunciation modern, but her particular style of *seva* is gendered. She, like other women renouncers I met, was approached as a mother. Baiji was a mother to the young Brahmin priests living in her ashram, and her concern went beyond their religious education. She would exhort them to keep their rooms clean, and lecture them on cleanliness and proper grooming. One evening, she reminded them that soap powder is always available for washing clothes and that she would teach them if they did not know how to do it properly. Also, as I have argued elsewhere (Khandelwal 2004:186–187), food is an area in which *sannyasa* is gendered. Procreative sexuality and feeding are realms that mark Hindu femininity (Marglin 1985). *Sannyasinis* renounce the former, but generally embrace the role of providing food that is the marker of motherhood. All holy people distribute food as *prasad*. However, for many women renouncers, food itself—its quantity, taste, and nutritional value—is a matter of such concern that feeding people is more than simply the distribution of blessing in the form of *prasad*. Baiji wanted those in her ashram to enjoy their food and to eat until satiated. This and her habit of eating last (more typically a guru would eat before disciples) were appreciated as expressions of her *seva* and maternal love.

Seva was the one word that Baiji's disciples consistently used to describe their guru's religious path, and she would inevitably begin our interviews by asking how she could serve me. That this is an inversion of classical ideas of the relationship between guru and householders or disciples, according to which the guru is the recipient rather than the giver of *seva*, was frequently pointed out by Baiji's disciples as one quality that distinguished her from other, more ordinary gurus. My conversations with Baiji and my observations about the details of ashram life illustrate the challenges that arise for a renouncer committed to the service of householders. While Baiji remained always calm as she directed and participated in ashram activities, she also reflected on these tensions as she recited the verse about a sadhu leaving it all, advised the young *pandit* about the dangers of serving householders, and returned periodically to her Himalayan ashram and its more contemplative life. According to her devotees, Baiji's spiritual knowledge and power enable her to consult deities directly, guide disciples, treat illnesses, administer charitable projects, and perform the humble work of running an ashram, and this capacity for *seva* is in turn taken as a sign of her saintliness and power as a guru. At the same time, for Baiji, the very act of being a guru creates tensions between the goals of service (*seva*), detachment (*vairagya*), and the destruction of ego (*ahamkar*). Egotism, most renouncers agree, is the most tenacious obstacle in spiritual progress, especially for those who take on the role of guru. Baiji herself acknowledged this from time to time; it would be much easier, she once said, to live a quiet life of solitude and contemplation and at times it seemed she longed for such a life.

Notes

1. The Dashnami renunciant orders, of which Saraswati is one, refers to the monastic orders ostensibly established by Shankara in the eighth century (but see Hacker 1964:29). This monastic federation is divided into ten ascetic lineages, which have separate names and practices, but all trace their spiritual descent from Shankara (Dazey 1990:284).
2. I spent several months with Baiji during 1989–1991 while conducting doctoral research and met her again on briefer visits to Haridwar in 1999 and 2005.
3. Baiji's disciples often discussed this in terms of the parallel distinction between *nirakar* (without form) and *sakar* (with form) forms of religiosity.
4. The earliest body of Indian literature is known as the Vedas, and religion of the Vedic period (approximately 1500–600 BCE) centered on the performance of elaborate ritual fire sacrifices by highly trained Brahmin priests. Toward the end of the Vedic period, notes Embree (1988), texts begin to appear in which Vedic ritual was attributed with symbolic meaning, and understanding this mystical meaning became even more important than actually performing the rite. These texts, collectively called the Upanishads, reflect the perspective of sages and ascetics rather than priests, for they hold out the possibility of enlightenment through ascetic practices. As the final stage in the development of Vedic philosophy, the Upanishads represent the "end of the Vedas" (Vedanta), so later philosophical schools of classical Hinduism that are based on the Upanishads are referred to as Vedanta (29). The Upanishads, in claiming the superiority of ascetic practices over ritual sacrifice, provide the primary textual authority for contemporary renunciation.
5. The term "*avadhut*" (literally, "one who has dispelled all imperfections") is used for any ascetic believed to be above all rules, although it is most appropriately used to refer to advanced Dashnami Nagas (Ghurye 1964:77).

6. When I visited Haridwar in 1999, I learned that Baiji spent most of her time in Uttarkhand and came to Haridwar only a couple of times each year.
7. Ramayana is the popular Hindu epic in which the hero Rama, incarnation of Vishnu, defeats the demon Ravana to rescue his beloved wife Sita.
8. The Vedas, as already noted, constitute the earliest body of Indian literature. The Vedas are associated with the dominance of the priestly class and are primarily concerned with ritual and ascetic practice.
9. Ramakrishna was a nineteenth century charismatic mystic who was barely literate in his mother tongue of Bengali. His states of rapturous devotion to the goddess Kali interfered with his abilities to perform his ritual duties, and for this reason he lost his job as a temple priest (Sinclair-Brull 1997:14). Although he was reportedly initiated by a Dashnami *sannyasi*, he continued to live with his wife and family and to wear ordinary clothes. Eventually, he began to attract disciples, especially young men from the educated Bengali middle classes who found his message about the limits of science and rational knowledge appealing (Chatterjee 1992:51). After Ramakrishna's death in 1886, 11 of his disciples donned the ochre robes of *sannyasa*, although the legitimacy of their *sannyasa* is suspect in the eyes of many orthodox renouncers and lay persons (Sinclair-Brull 1997:24).

Bibliography

Beckerlegge, Gwilym. 1998. Swami Vivekananda and Seva: Taking "Social Service" Seriously. In *Swami Vivekananda and the Modernization of Hinduism*. William Radice, ed., pp.158–193. Delhi: Oxford University Press.

Chatterjee, Partha. 1992. A Religion of Urban Domesticity: Sri Ramakrishna and the Calcutta Middle Class. In *Subaltern Studies*. Vol. VII. Partha Chatterjee and Gyanendra Pandey, eds., pp. 40–68. Delhi: Oxford University Press.

Dazey, Wade H. 1990. Tradition and Modernization in the Organization of the Dasanami Samnyasins. In *Monastic Life in the Christian and Hindu Traditions: A Comparative Study*. Austin B. Creel and Vasudha Narayanan, eds., pp. 281–321. Lewiston, NY: Edwin Mellen Press.

Embree, Ainslie T. 1988. *Sources of Indian Tradition*. Vol. 1. *From the Beginning to 1800*. Revised edition. New York: Columbia University Press.

Ghurye, G.S. 1964. *Indian Sadhus*. Bombay: Popular Prakashan.

Hacker, Paul. 1964. *On Sankara and Advaitism*. Reprinted in Wilhelm Halbfass, ed. *Philology and Confrontation: Paul Hacker on Traditional and Modern Vedanta*, pp. 27–32. Albany: State University of New York Press.

Khandelwal, Meena. 2004. *Women in Ochre Robes: Gendering Hindu Renunciation*. Albany: State University of New York Press.

King, Ursula. 1980. Who is the Ideal Karmayogin? *Religion* 10 (Spring):41–59.

Marglin, Frederique A. 1985. *Wives of the God-King*. Oxford: Oxford University Press.

Narayan, Kirin. 1993. Refractions of the Field at Home: American Representations of Hindu Holy Men in the 19th and 20th Centuries. *Cultural Anthropology* 8(4):476–509.

Sinclair-Brull, Wendy. 1997. *Female Ascetics: Hierarchy and Purity in an Indian Religious Movement*. Surrey: Curzon Press.

Chapter 3

Living Practical Dharma: A Tribute to Chomo Khandru and the Bonpo Women of Lubra Village, Mustang, Nepal

Sara Shneiderman

Introduction

Chomo Khandru was 80 years old when we met. I was 20. She welcomed me into her home for one night, which soon turned into many. It was 1995, and I had traveled to the ethnically Tibetan area of the Nepal Himalayas known as Mustang to conduct research on the lives of *chomo*—female religious practitioners of the Tibetan Buddhist and Bon traditions who live independently in villages, without the support of an institutional monastic setting.[1] Chomo Khandru was the oldest woman in Lubra, a tiny settlement of 14 houses sheltered in a side canyon of the mighty Kali Gandaki river, and she was arguably the senior most *chomo* in the entire area (figure 3.1). She was an old woman with much to teach and no disciples, and I was an eager young student. As I gained her trust, the story of her life as both a celibate ascetic and a worldly trader on the historic Tibet to India salt-grain trading trail unfolded. With an unusually sharp memory for dates, names, and places, and a down-to-earth way of explaining abstract dharmic (religious) concepts, she was a gifted raconteur and teacher. It became clear that the experience of telling her story and explaining her beliefs was important to

Figure 3.1 Portrait of Chomo Khandru, 1998
Source: Photo by Sara Shneiderman.

her, just as the experience of listening was to me. Together we entered the special space of transmission. I use the word in its spiritual sense, where it implies the ritual passing of knowledge and practice from one generation to another.[2]

Since that first meeting, I have puzzled over how to tell Chomo Khandru's story, an individual life story that recalls the rich collective history of women's religious practice in the village of Lubra, as well as the larger Himalayan region of Mustang, and the countries of Nepal and Tibet. Her story also foreshadows the cultural, political, and economic challenges that Chomo Khandru's descendants now face as Tibetan religion becomes increasingly globalized. When Chomo Khandru died in November 2002 at the age of 87, a chapter of women's history went with her, and my desire to write about her life grew. Here, I pay tribute to her, as well as to the generations of women who came before her, and to those who will come after. To do so, I move back and forth between the narrative of Chomo Khandru's individual life and a broader analysis of the social dynamics surrounding women's religious practice in the region. Rather than painting a dualistic picture of tradition pitted against modernity, or valorizing Tibetan women's agency in the past while viewing contemporary women as powerless victims, I hope to illuminate the complex webs of spiritual accomplishment and frustration, physical hardship and comfort, gendered and non gendered experience that have shaped women's religious lives in Lubra over time.[3]

Janice Willis has described women's roles within the Tibetan cultural milieu as "dynamic, bustling, diverse, and fluid" (1989:100). In this chapter I focus on the tension between such fluidity and fixity in constituting women's religious identities in Lubra. In Chomo Khandru's life, this dynamic manifested itself as an opposition between the dharma of action, or "practical dharma," that she saw as her own primary mode of spiritual engagement, and the dharma of study, which she saw as the domain of men within institutional monastic contexts. This opposition was not hard and fast, as Chomo Khandru did have some access to informal religious education, but she saw its role as secondary to meditation and ritual practice in her spiritual life. I suggest that over the second half of the twentieth century, during which Chomo Khandru lived most of her adult life, this local conception of fluid religious identities began to come into conversation with a more static notion of religious identities that has been inadvertently cultivated by several extra-local entities, including the Tibetan monastic establishment in exile, the Nepali state, and foreign charity and development organizations.[4]

All of these players emphasize models of religious authenticity based on institutional power and formal learning, whereas local models historically placed a higher value on noninstitutional forms of individual spiritual accomplishment and hereditary religious authority.[5] Idealized extra-local notions of how Tibetan religion "should" be structured have encouraged shifts in local conceptions of religious authority.[6] These trends tend to marginalize the nonmonastic, nonliterate modes of religious practice that were historically most accessible to women in many Himalayan areas. As an example of the flexibility that female practitioners experienced, during Chomo Khandru's youth, celibacy for *chomo* was valued but not strictly enforced. Although both men's and women's potential avenues for spiritual practice have become less flexible in reaction to the ongoing engagement between these varied forces, I argue that women in particular have found it more difficult to follow in Chomo Khandru's footsteps to lead renunciant lives, despite the widespread rhetorical commitment to improving women's education and economic status on the part of political and religious institutions over the past several decades.

Many of these difficulties have to do with the conflicting sets of social and spiritual expectations that female practitioners face: quite often, the social roles that they are expected to play as women do not allow the development of the qualities that they are expected to demonstrate as spiritually potent individuals. In large part, this disjuncture is related to the increasing value placed on reading and writing as the exclusive keys to religious development, skills that are gendered as male in the Tibetan context, in contrast to female-gendered skills such as weaving. Chomo Khandru moved back and forth between these domains with a gender-bending fluidity in a way that is increasingly uncommon for today's women.

Locating Lubra and the Bon Tradition

Lubra is located in the lower half of Mustang District, Dhaulagiri Zone, Nepal. Part of the larger 19-village community known as Baragaun that stretches along the Kali Gandaki river valley, Lubra is just out of range of the heavily trekked Annapurna Circuit trail (figure 3.2).[7] In this sense, it straddles two worlds: it remains relatively isolated and tourist-free in comparison to other Baragaun villages, but its proximity to them furnishes access to the amenities, trade opportunities, and ideas that tourism and development

Figure 3.2 View of Lubra Village, 1995

Source: Photo by Sara Shneiderman.

projects have brought to the area. Villagers depend upon a mixture of subsistence agriculture and cash income from seasonal trade for their livelihoods.

The inhabitants of Lubra's 14 houses (numbering approximately 80 men, women, and children) are descendants of the twelfth century Lama Tashi Gyalzen, who ventured into the Mustang area from Tibet in search of a spiritual teacher believed to reside there (Ramble 1983). Upon reaching Lubra, he founded a lineage of noncelibate householder priests that continues today (Ramble 1984). Lama Tashi Gyalzen appears to fit the description of the "Himalayan frontier lamas" that Joanne Watkins (1996:200) has described in her discussion of women's religious roles in Manang, the Nepali administrative district that lies to the east of Mustang. Watkins suggests that such eclectic individuals left the rigid religious and political hierarchies found in central Tibet to establish noninstitutional forms of practice in Himalayan borderland areas. Among other things, the religious traditions that these lamas founded tended to enable greater religious participation for women because they emphasized practical aspects of spiritual development, which could be undertaken in the home, rather than formal study that required travel to monastic institutions elsewhere.

Tashi Gyalzen was a follower of the Bon tradition, and the villagers of Lubra maintain that identity today.[8] This line of Tibetan religious practice has a long and rather unclear history that has defied clear categorization by cultural historians. Bon's adherents claim that it preceded Buddhism as the earliest form of organized Tibetan religion and that the historical figure of Shakyamuni Buddha was a later incarnation of Bon's founder, Sherab Miwo. Bon cosmogony diverges from its Buddhist counterpart in terms of its primary deities and origin stories.[9]

Early Western scholars often tended to reify and legitimize elite Tibetan perspectives that saw Bon as a deviation from "orthodox" Buddhism. For this reason, the Bon tradition often received negative and inaccurate treatment in scholarly works. Geoffrey Samuel offers an explanation for this tendency:

> Bon remained a kind of amalgam of early Tibetan religion, contemporary Tibetan folk religion, black magic and sorcery, a generic label for all the aspects of Tibetan religion which did not fit neatly into Western stereotypes of proper Buddhism. The real problem with this approach is that it collapses a very complex historical process, in which Tibetan Buddhism and the Tibetan Bon religion developed side by side, into an unhistorical model in which pure Buddhism comes from India and degenerates under the influence of the native Bon religion. (1993:323)

Due to prejudices like these, Bon was long marginalized within both the study of Tibetan religion and within contemporary Tibetan cultural politics in exile (Cech 1993). However, the Dalai Lama has now acknowledged Bon as a fifth sect of Buddhism, and the Bon tradition's high lamas are well respected by male and female Bonpos, and also by Buddhist monks and laity.[10] Noting these similarities, Samuel suggests that "the modern Bonpo are to all intents and purposes the followers of a Buddhist religious tradition, with certain differences of vocabulary from the other four major traditions of Tibetan Buddhism, but no major difference in content" (326).

In some ways, Lubra's Bon-ness contributes to a sense of "double oppression" for women like Chomo Khandru, who are not only women but also members of a once marginalized religious sect. Lubra is the only entirely Bon village in Mustang (although there are individual families scattered elsewhere), and this means that there are far fewer alternatives available to Lubra's women for formal religious practice, since all of the nunneries in the area belong to either the Sakya, Kagyu, or Nyingma Buddhist sects. This is one of the reasons why women's religious practice in Lubra has always been particularly fluid and noninstitutionalized: with the closest Bon monasteries located in eastern Tibet before 1959, and in Kathmandu and northern India afterward, women had little access and were therefore compelled to develop their own forms of spiritual practice at home. At the same time, however, Lubra women describe Buddhism and Bon as *experientially* similar, noting only superficial differences such as the direction of circumambulation or the particular words used in mantras. Krystyna Cech has noted that for many Bonpos, "Identity is a fluid rather than a rigid concept; it allows for the construction, deconstruction and reconstruction of its aspects" (42). Fluidity is therefore a hallmark of religious identity for both men and women in Lubra, both in terms of the flexibility between Buddhist and Bon identifications and in terms of the particular modes of practice that are available to them. Indeed, many of Lubra's women marry into Lubra from the surrounding Buddhist villages and have therefore experienced both religious traditions. So although Chomo Khandru's life story is first that of a Bonpo woman, I believe it resonates strongly with the experiences of Buddhist women from Mustang and other Himalayan areas as well.

Life as a Chomo: Birthright or Burden?

As we first began to talk, Chomo Khandru had a difficult time understanding my interest in her as a *chomo*. She would direct me to the male village lamas whenever I inquired about her religious practice.[11] Soon I began to realize that instead of asking questions that focused on her religious activities as a *chomo*, assuming that this was the primary feature of her self-identity, I did better to ask general questions about her life experiences that did not make her feel self-conscious about her lack of formal religious training. The answers to these questions reflected the virtues of the "practical dharma" that Chomo Khandru was initially embarrassed to admit that she practiced. This English phrase best captures a concept that Chomo Khandru and other village women described in various Tibetan and Nepali terms at different times. In short, they contrasted the "dharma of action" with the "dharma of study," associating themselves with the former, which they saw as a lower status mode of religious practice, and associating men with the latter, higher status mode. This notion of "practical dharma" resonates with the *kriya yoga* that Hausner notes and also the *seva* that Khandelwal describes for Hindu women renouncers (this volume).

Born into a family with seven daughters and no sons, Chomo Khandru became a *chomo* at the age of 11. In Lubra, she explained, this was the fate of all second daughters in families with three or more daughters.[12] As in Zangskar, which Gutschow describes elsewhere in this volume, various economic considerations compelled families in Lubra to offer their daughters as *chomo*, although unlike in Lubra, in Zangskar there was no strict rule by which every middle daughter became a *chomo*. In both places, first daughters were destined for early marriage, and in certain situations they could also inherit property. For the people of Lubra, requiring second daughters to become *chomo* ensured that the family would have at least one member focused on generating religious merit, while also creating a time lag between the marriages of the first and third or younger daughters, which enabled the family to save money in between for later dowries and rituals. The local marriage system relied upon fraternal polyandry—meaning that a single woman married two or more brothers simultaneously—in order to maintain the integrity of small plots of land as they were inherited by each generation through the male line.[13] Marriage prospects for many women were therefore severely limited in this small community, and a tradition of female renunciation was a good solution to the problem of too many unmarried women. Parents could gain religious merit by pledging their daughters as *chomo*, yet maintain the right to their labor rather than lose it to a prospective husband's family. Many middle daughters slated to become *chomo* cursed their fate for keeping them from married life—it was the rare few like Chomo Khandru who considered themselves fortunate in gaining access to the basic religious training and spiritual practice usually reserved for men.

Chomo Khandru's identity as a dharma practitioner was confirmed through a hair-cutting ceremony that took place at the village *gompa* (temple), during which she and her family made offerings to a visiting high lama and she formally pledged to devote herself to religious life.[14] However, it is unclear whether this ceremony included formal vows. In any case, such details seemed superfluous, since the community as a whole was well aware of her role and would treat her as a *chomo* whether she had been ordained or not. From that point on, Chomo Khandru began an informal process of religious training with two well-respected older *chomo*, Tsultrim and Dawa Kunzum. Educated by Yeshe Gyalzen, a Tibetan *tulku* (reincarnate lama) from the eastern Tibetan region of Kham who had spent long periods of time in Lubra during these *chomos'* youth, both women were highly literate in Tibetan and accomplished in meditative practices. Chomo Khandru and other Lubra women spoke of Tsultrim and Dawa Kunzum with great respect: they were the only literate women of their generation and the only village women who undertook extended periods of solitary meditation. These *chomo* passed on their Tibetan reading and writing skills to Chomo Khandru, but for various reasons, she felt that she was never able to develop them to the high degree that her mentors had.[15]

Although Chomo Khandru had a great desire to engage in spiritual practice, she was instead compelled to work as a worldly trader for many years. Since there were no sons in the family, and she was the only daughter without responsibilities to her own husband and children, she became the heir to her father's trading business. Instead of fostering her religious practice, her identity as a celibate *chomo* required her to serve as the family's missing man, devoting the time she would have spent on religious practice to breadwinning work for the family.

One of her primary responsibilities was overseeing the family's concerns in the salt, wool, and grain trade that Himalayan communities have traditionally relied upon.[16] From about 1935 to 1959, when Chomo Khandru was a young woman between the age of 20 and 45, she worked as a trader. She traveled north to Lo Monthang (the walled city that served as the seat of Mustang's king) at least once a year, carrying Lubra's wheat to trade for Tibetan salt and wool. She would travel with her male relatives and a train of grain-laden *dzo* (yak/cow crossbreeds). Rarely did any other women accompany them. While trading, she was often frightened that men would attack her and abduct her into a forced marriage, thereby compelling her to give up her religious celibacy. In order to protect herself, she dressed as a man, and with her hair already short as a *chomo*, she could usually pass as one. She carried a *khukuri*, a traditional Nepali knife, by her side and took turns with her male relatives keeping guard over their hard-won grain. "Sometimes, I was challenged by groups of drunk men we met on the trail," she explained. "They wanted to know whether I was male or female. They wanted me to show them! Of course I didn't!" she chuckled. "I had to wave my *khukuri* to keep them away!" Sleeping out in the open among men who were away from their wives for months at a time, she struggled to maintain her celibacy and the religious honor that went along with it.

Not all *chomo* remained celibate, but her personal commitment to this lifestyle was a source of pride for Chomo Khandru. She was the only *chomo* of her generation who had not given birth, but she respected the others for their choices. Her attitude toward celibacy became more sharply defined in her old age, and seemed to highlight a broader social shift on this issue. Due to the influences of both normative Hindu ideals encouraged by the Nepali educational system, and the development of Buddhist monasticism, by the late 1990s it was becoming less and less acceptable for women (*chomo* or lay) to bear children out of wedlock, and Lubra's villagers were beginning to view celibacy for both men and women as a prerequisite to serious religious practice.

Due to the constant fear that her subterfuge would be discovered and the long periods of time she spent away from Lubra, the trading years were not a happy period in Chomo Khandru's life. At the same time, she was grateful for the freedom from domestic responsibilities that allowed her to travel and visit pilgrimage sites that would have been out of range for most Lubra women, who are generally responsible for maintaining the "home" to which traveling men return.[17] Although she traveled as a businesswoman rather than as a religious mendicant, Chomo Khandru's journeys challenged the existing expectations of women's roles in Lubra. In this sense, her wanderings were not unlike those Hausner describes for female Hindu renouncers elsewhere in this volume. Like those *yoginis*, Chomo Khandru had mixed feelings about the traveling life: on the one hand it allowed her to engage independently with a world that most Lubra women never knew, but on the other hand the constant movement kept her from deepening her spiritual practice in a consistent manner. If Chomo Khandru could have had her way, she often told me, she would have stayed at home. But I always wondered if doing so would have undercut the very foundations on which her reputation as a spiritual practitioner were built—had she stayed at home, she would have been seen as just another woman, without the special, "male" qualities gained through traveling, which gave her the necessary credentials to be well respected as a religious figure in the eyes of the village.

On Writing and Weaving: Gender-Bending Identities

Chomo Khandru's story demonstrates the fluidity of both gender and religious roles in Lubra. As for nuns in Zangskar (Gutschow 2004), the religious agency of *chomo* can be limited by their femininity—in many situations they do not have equal access to the formal religious training that their male counterparts receive. At the same time, Chomo Khandru's life history suggests that both gender and religious identities are mutable. The religious role of *chomo* can enable individuals to bend socially sanctioned gender rules. For example, it can enable a woman to become a symbolic male or, as in the case of Chomo Khandru, to take on certain qualities perceived as man-like. Anthropologist Charles Ramble writes that in Lubra's Tibetan dialect, "male honorifics are also used of celibate nuns from any rank" (1984:133). This suggests that Chomo Khandru's gender-bending trading activities were not aberrations; rather women who become *chomo* were perceived to take on male qualities in both the social and spiritual worlds. There is a specific kind of maleness that *chomo* must emulate if they are to command community respect. Ramble continues that although *chomo* should be addressed with male honorifics, "villagers confided to me their reluctance to address nuns in this way if they were not literate" (133). In fact, the male honorific marks literacy skills (traditionally in the Tibetan liturgical script) more than it marks physical gender features or a particular level of spiritual attainment; *chomo* are addressed as men because as religious practitioners they are expected to be literate. Reading and writing are generally considered to be male activities, and the fact that few women read Tibetan is one of the primary reasons that they are prohibited from maintaining a more active role in the village's literate religious life. Although young girls now learn Nepali in government schools, they are not proficient in the Tibetan "social script" (March 1984:737). Tibetan is the language that has traditionally encoded social and religious relations in Lubra, while Nepali is a "foreign" language of governance that belongs to a different language family (Indo-Aryan rather than Tibeto-Burman) and has a different alphabet. One lama made it clear to me that it was necessary to read Tibetan to become a full-fledged Bon practitioner; he continued to explain that women were barred from becoming lamas simply because they could not read texts, not because there was anything inherent in their feminine nature preventing them from taking on that role.

This conflation of gender, religious roles, and literary skills can create confusion at several levels: villagers told me that they sometimes had trouble knowing which pronoun to use when addressing a *chomo* whose level of literacy they did not know, and, conversely, they were confused by the fact that there was no easy way to address a literate woman, *chomo* or otherwise, without calling her a man. Again, these linguistic limitations suggest that *chomos'* perceived gender identities shift back and forth between "male" and "female"—although they are clearly women in the biological sense, their ability to read and write marks them with male-gendered qualities. For this reason, *chomo* who cannot read maintain the most ambiguous status: as religious practitioners, they should in theory be literate, therefore becoming honorary men, but without Tibetan literary skills, they are stranded between conflicting social expectations and gendered roles.

This state of affairs resonates with Kathryn March's discussion of writing and weaving as gender markers in Tamang society:

> For the Tamang, weaving and writing are not only technical skills but dense symbols of gender. They are gender symbols not only because they tell the Tamang about the separate roles of the sexes, but because they are about what transpires between the sexes as each defines the other. Two opposing conceptions of the world emerge as Tamang men and women view one another; gender symbolizes both the opposition and the reflexivity of these world views. (729)[18]

As in the Tamang community that March describes, in Lubra weaving stands in opposition to writing as the exclusive province of women. Women weave brightly colored blankets and aprons, the sale of which provides one of the village's few modes of cash income. Chomo Khandru never learned to weave, or perhaps more accurately, was never taught to weave by her mothers and sisters. Off trading as the "man" of the house, she was expected to write and read, but not to weave. Although she learned how to spin—an occupation that men and women share—Chomo Khandru's "male" identity as a trader and religious practitioner prohibited her from taking on the "female" skill of weaving. Speaking of a female figure prominent in Tamang myths, March notes that "weaving is associated . . . with the moment of her break with organized religion, her separation from her father, and the cloud of illegitimate pregnancy, all of which hampered her religious efforts" (733). Clearly, since weaving has similar connotations in Lubra, possessing the skill would only have complicated Chomo Khandru's already challenging position as a female religious practitioner.

One of the ironies of the equation made here between literacy, maleness, and the dharma of study (and its inverse: weaving, femaleness, and the dharma of action), is that the goal of orthodox Tibetan religious practice—both Buddhist and Bon—is to transcend intellectualized understandings of the self as objectively real to reach an experience of the ego as illusory and contingent. It is precisely this unintellectualized, embodied aspect of spiritual practice—the dharma of action—that laywomen most often engage in their daily lives. As one feminist observer of Tibetan religion has put it, "the *ideals* of these religious systems are often akin to the kinds of experiences women have in day-to-day life. This means that their experiences are perceived to fit in with the ideals of the religion, *even though these experiences are the ones they wish to change*" (Campbell 1996:156). In other words, although women are identified with the internal spiritual qualities (embodied in their weaving) that male-dominated religious traditions promote as ideals to attain, women themselves are excluded from the associated institutional structures of the religion precisely because they do not possess the learned skills (such as writing) on which the religion pragmatically relies.

Chomo Khandru and many of Lubra's laywomen seemed to be aware of this tension, although they did not speak about it in these specific terms. In statements to me, they always accorded their own spiritual attainments through "practical dharma" less value than those of men who engaged in the "dharma of study." Yet they still believed strongly in the importance of "practical dharma," precisely because it was practiced as part of daily life in the village and therefore generated direct, immediate benefits relevant to them as women who spent most of their time at home. This was in contrast to the ostensibly greater, but abstract and unquantifiable, merit generated by monks in far-away institutional settings.

For instance, I was told by several people that one of the reasons there was no strict rule of celibacy for Lubra's *chomo* was because having the experience of childbearing and rearing enabled *them* to empathize more fully with laywomen's experiences, and therefore to meditate more fully on motherhood as an experience of great compassion as one plank of their spiritual practice. Tsultrim and Dawa Kunzum—the two literate *chomo* who taught Chomo Khandru—both bore illegitimate children (neither of them married, although Tsultrim was the long-term partner of Lama Yeshe Gyalzen), but this did not compromise their reputation as spiritual practitioners. If anything, motherhood strengthened their ability to engage with dharma at the practical level and serve as role models and teachers to other women. They inhabited an intermediary space between male and female identities: they could read and write like men, yet they never received more formal education or institutional support because they were women. When at home they lived and worked with their families and cared for children like women, yet they periodically left for pilgrimage or business trips that took them away like men. This intermediate position was socially acceptable in the village context of Lubra, allowing them to maintain a fluid identity that did not depend upon institutional legitimization.

Changing Religious, Economic, and Educational Landscapes

Chomo Khandru's trading work came to an unexpected halt when the Nepal-Tibet border was sealed after the Chinese occupation of Tibet in

1959.[19] This political shift allowed her to devote the latter part of her life to the religious practice that had eluded her in youth. Finally, she had time to read texts and sit in meditation. Of course, she still had work to do—she became an indispensable help to her sister Palsang, who was raising six children. When I met her, she was still working alongside her sister, serving as a second grandmother to her nephew's two small sons and doing a full share of field work and housework. Domestic labor dominated her days—she once said sarcastically in response to a question about why she had not engaged in more formal religious training: "Every day we have work. Only when we die, we don't have work!"

Nevertheless, Chomo Khandru could work hard at home or in the fields during the day and spend her evenings reading by flickering candle light. Through her obvious devotion to the dharma, she became a highly respected village figure, the only woman ever called to conduct important rituals in the village temple or individual homes with the village lamas. She memorized basic religious texts and spent the little free time she had meditating, but she was never able to study at the more advanced philosophical level that she desired. By the time she was 50, her eyesight was beginning to fail. At that time, no one wore eyeglasses or knew of any other cure. Slowly, over the next ten years, the world receded into a fog, and she could no longer study new texts. She had committed the most important works to memory and could still chant them with the lamas when she was called. But slowly these too faded. "My dreams of becoming enlightened are still just dreams. I am old now, and cannot undo all of the bad karma I have accrued in this life. It is too late," she said remorsefully one evening as we sat around the fire. Although venerated by the villagers for her age and experience, her prayers were no longer thought to be as effective as those offered by younger monks trained formally in the new monastic centers of India and Kathmandu that Chomo Khandru had never visited.

Such institutions began to develop after the Tibetan flight into exile after 1959, often with support from the growing numbers of Western practitioners of Tibetan Buddhism and Bon.[20] Although the people of Lubra were citizens of Nepal, the political shifts north of the border had severe consequences for them. The border between Tibet and Nepal was closed, and the itinerant lamas who had provided the focal point of Lubra's religious life no longer came. As the Dalai Lama and many other Tibetans fled into exile in India, the center of religious authority for the people of Lubra and other Himalayan villages suddenly shifted from Tibet to the exile towns of India and Nepal, where many lamas reestablished their monasteries. The people of Lubra had never previously needed to visit monasteries in Tibet; the lamas had always come to them. Now they were bypassed. The exiled Tibetans had lost their village context and began to build monastic institutions that were not constrained by local politics and had little place for the noninstitutional traditions of village practice. At the same time, Lubra became ever more constrained by its out-of-the-way location. Ironically, the transnational development of Tibetan religion seemed to have made Lubra more isolated than ever before from the new religious and cultural centers.

This shift had grave effects for the women of Lubra. No longer could they receive religious teaching from high lamas at home. Instead, they had to travel to far-away places such as Dharamsala (the Dalai Lama's seat in Northern India) or Kathmandu. While boys were now sent off for study, girls rarely were; they were expected to stay at home to provide labor for their families. Menri Gompa, which had been the most important Bon *gompa* in Tibet, was reestablished in 1969 as the seat of the Bon tradition in exile at Dolanji, in Shimla, Himachal Pradesh, India (Skorupski 1981). Funded in part by foreign donations, the Dolanji monastery began offering full scholarships for studies up to grade eight to children from Himalayan villages like Lubra. Although in theory these scholarships were open to boys and girls, parents rarely felt able to send both their daughters and sons, and ultimately boys received priority.

Instead, girls were left at home and sent to study at the Nepali government school established in the village in the late 1980s. Although this was certainly a positive step forward in the overall educational level of village women, the *type* of education had unexpected effects on women's abilities and motivations for religious practice. Most importantly, although girls gained literacy in Nepali, this opportunity in fact curtailed their ability to learn the Tibetan they would need for serious religious practice. With the time limitations placed on girls by their domestic responsibilities, they could barely manage to study one written language, let alone two. Furthermore, the school provided a false sense of educational security, and the one lama who had previously tried to provide some education in Tibetan for village girls along with Chomo Khandru ceased these activities after the establishment of the school. Most importantly, the education in Nepali provided younger women with access to the world of trade and commerce. To the dismay of many older villagers, many girls, including those intended to be *chomo*, lost their motivation for religious practice and left the village to become businesswomen elsewhere.

From Chomo Khandru's perspective, Lubra's younger generation of women had been converted to the "religion of money," a change that in her view was in large part due to the emergence of secular Nepali language education in the vacuum left after the itinerant lamas stopped coming to teach in 1959. However, Chomo Khandru herself had spent years as a trader and was not untainted by worldly concerns. So what was her critique of the young women? She felt that her own engagement in business was not out of choice but of necessity, whereas the younger women wanted nothing more than to leave the village for commercial pursuits. This highlights again the tension between traveling and staying in one place, discussed earlier. Chomo Khandru claimed that in her time, trade was a necessary evil that she was pushed into by her family for economic survival, while she would have chosen to stay home to focus on spiritual practice if given the option. At the same time, as mentioned above, traveling did afford Chomo Khandru a certain freedom to visit pilgrimage sites that otherwise would have remained inaccessible to her as a woman. In contrast, the women of her great-niece's generation had no interest in dharma and saw traveling and trading as ends in themselves, often leaving the village against the wishes of their families. Also, they traveled southward to the Nepali city of Pokhara and onward to India, rather than northward to Tibet. This shift was in large part due to the disappearance of northern trade opportunities after 1959, but there was no question that Nepali education had also contributed to shifting the younger generation's orientation permanently southward.

At the end of my first stay with Chomo Khandru in 1995, she told me about the dream she shared with many other villagers: to revitalize Lubra by rebuilding the village's historic *gompa*, founded in the twelfth century by Lama Tashi Gyalzen, as a center for religious education that would provide a Bon education for *both* boys and girls. This would reintroduce Tibetan literary skills as an important educational focus, and it would help stem the brain drain—or dharma drain—that the village was experiencing.

Ideas Become Action: Building a Bon School

By the summer of 2001, this dream was on the way to realization, although in a somewhat different form than Chomo Khandru had imagined. A Western visitor had secured funding from a development agency to rebuild the *gompa*, and by late 1997, the old *gompa* was usable once again for ritual practice, although it had not been developed for educational purposes. In 1999, funding was secured to build a separate Bon school named Chasey Kengtse Bon Hostel. The plan was that the hostel would work in tandem with the Nepali government school to provide morning and evening classes in Tibetan language and religion. The donor organization was a small foreign charity run by several women devoted to advancing education, with a particular emphasis on girls' education. By the summer of 2001, the mammoth school building—with 14 toilets, in a village of 14 houses that previously had no toilets—had been constructed just below the rebuilt *gompa*. However, disagreements over how the school should operate prevented it from opening until 2002. Even after its opening, infrastructural and logistical problems plagued the school, culminating in the tragic death of one child in early 2004 after part of the school building collapsed. Far from providing the idealized vision of Bon religious education, the project became the terrain for a divisive debate about religious authenticity, which threatened to in fact curtail rather than encourage women's religious and cultural agency in the village.

Much of the difficulty arose due to confusion among the donor organization's board of trustees about the role of women in Tibetan societies and how best to encourage their participation. Familiar only with the highly literate, monastic model of Tibetan religion that was emphasized as Buddhism and Bon moved West, the trustees of the donor organization assumed that developing a school built upon such a model would be the most effective way to encourage girls' education and religious participation. This approach failed to acknowledge the local prevalence of nonformal education for female children, which generated noninstitutional forms of religious agency, such as that which women like Chomo Khandru exercised in Lubra before 1959. Development programs that aim to increase women's access to religious education in places like Lubra must acknowledge the fluidity of women's traditional religious roles in the local context and work to maintain such options, rather than promoting participation in male-dominated, newly popular institutional modes of religious practice as the only way forward for women.

This clash between a model of religious fluidity and one of monastic education is behind the drama of the Bon school in Lubra. The new Chasey Kengtse Bon Hostel was designed to replicate in Lubra the monastic structure of education at Menri Gompa in Dolanji, thereby introducing an altogether new form of religious institution to the village. The masterminds behind this plan were the three trustees of a Nepali NGO, Mustang Bon Action, which took on the role of liaison between the foreign donor organization and the village school committee. The Mustang Bon Action board of trustees consisted of one concerned Western visitor to Lubra and two businessmen based in Kathmandu. Both men had studied at Dolanji themselves, and they were convinced that the institutional monastic model was the only one that could effect the religious revitalization that Lubra's residents desired. Yet neither of them had spent any of their adult life in Lubra, and they had dismissed alternate models for religious education that might have been more in line with the village's lived traditions. Although some villagers had misgivings from the start, many had been convinced by the trustees' argument that they needed a monastic-style educational institution in order to maintain their religious tradition in the future. These villagers had visited Dolanji and other such institutions and were impressed by the sheer physical grandeur and the religious power centralized there, both largely thanks to Western financial support. Identifying the monastic aspects of the tradition as the most "genuine," Western students of Tibetan lamas and aid agencies have donated large amounts of money to such institutions, while until now relatively little has gone toward maintaining local religious traditions in areas like Mustang (Childs 2004; Moran 2004: chapter 4). By the late 1990s, many people in Lubra were very conscious of these inequalities, and they had begun to view their own, non-monastic modes of practice as backward, looking toward the new Western-supported monasticism as the highest source of spiritual authority. As one of Lubra's older lamas said of his son who was studying at Dolanji for the *geshe* degree—the equivalent of a PhD within the Tibetan philosophical system—"He will be a much higher lama than I have been, because he is a monk, while I am married." This is additional evidence of the shifting attitudes toward celibacy, as discussed above.

In addition, many villagers thought that the foreign donors who had funded the hostel construction would be dissatisfied if the result was not a grand building on par with Dolanji and the other large monasteries that foreigners were likely to be familiar with. In private, many villagers confessed that they would prefer a smaller building in keeping with existing local architecture, which would then free up more funds for teacher training and curricular development. However, these individuals felt that proposing their ideas to the charity was impossible, because this would expose Lubra's inability to live up to what the villagers perceived as foreign expectations of Tibetan religion. Western support for, and romanticization of, Tibetan religion was transforming local conceptions of religious authority, even in the very act of providing support for "traditional" monastic training. This created a dynamic that marginalized the nonmonastic, nonliterate modes of religious practice that had historically been women's forte in many Himalayan regions.

So although the villagers were in theory eager to adopt the monastic model proposed by the trustees, in practice, this model often conflicted with their own priorities. Villagers became increasingly dissatisfied with the project as the school building neared completion and value-laden decisions about the school's curriculum and admission criteria had to be made. A series of disputes arose between the trustees of Mustang Bon Action and the village school committee around how the school should be operated, most of which had to do with competing notions of religious authority: for example, should the head teacher be a local married *tulku* who had not been trained in a monastery, yet commanded respect from many villagers, or a young celibate monk with credentials? The trustees were eager to promote practices and structures they saw as essential to maintaining "Bon culture," but which in reality had never been part of life previously in this historically Bon village. Such "inventions of tradition" (Hobsbawm and Ranger 1983) served to alter village lifestyles in unintended ways, rather than preserving existing cultural practices.

A case in point was the issue of uniforms: the trustees wanted the students to wear a complicated traditional Tibetan costume with supposedly "Bon" colors and designs for the purpose of "preserving Bon culture," yet the villagers had never worn such costumes themselves. Instead, they saw such uniforms as an unnecessary expense that could prevent the poorest children of the village from attending school. They also worried that unusual uniforms might in fact alienate their children from the Nepali world around them, rather than giving them a positive sense of Bonpo identity within it, as the trustees had envisioned.

The biggest invention of all was the notion that the only authentic form of religious education was that provided by monastic-style institutions, through the "dharma of study." In the eyes of the Mustang Bon Action NGO's trustees, the "dharma of action" that Chomo Khandru had practiced was not a form of spiritual practice worth transmitting to future generations because it unfolded in the nooks and crannies of daily life, in unpredictable patterns that did not adhere to the principles of monastic discipline. Moreover, it was not structured around textual knowledge. When Chomo Khandru approached the trustees with several suggestions—including a request that children not be required to live in the boarding hostel but instead be sent home at night so they could be with their families—they were dismissed as the follies of an old woman, and she was told that the children would never learn anything properly if they did not adhere to strict disciplinary rules.

Chomo Khandru was not the only woman denied a role in shaping the future of the school. There were no women among the seven members of the village school committee. According to the trustees, there were no women who had enough knowledge of what religious education should be like to participate in the process of shaping it in the village. Village women on the whole were skeptical of the project, and they felt uncomfortable with the monastic model advanced by the trustees. Having never known religious women apart from the noncelibate *chomo* who always lived in the village, rather than in a monastic setting, many women equated the institutional model with men and feared that their daughters would be discriminated against at the school. As one lay woman put it, "Why will it benefit my daughter to go and sit there all day if they only pay attention to the boys? She could be doing better work at home." The foreign donor organization's trustees could not understand why women were not more enthusiastic about the project.

In short, by challenging the traditional fluidity of gender and religious identities with the introduction of an educational institution modeled on Tibetan exile monasteries, both the foreign sponsors of the school and the Kathmandu-based trustees had overlooked the ways in which Lubra's women might be best served. As a symbol of the monastic model of religion, which until recently had remained peripheral to the villagers' own practice, the project imposed a fixed notion of how religious identity should be constructed. In the process, rather than being valued as knowledgeable advisors, elders like Chomo Khandru were cast as unfortunate reminders of the village's "backward" religious past and relegated to the margins of what was constructed as its "modern," monastic future.

Continuing the Lineage

Over the last ten years (1995–2005), I have returned to Lubra many times. In 1999, I became *mithini* (Nepali), or *drogmo* (Tibetan)—"ritual sister"—with Chomo Khandru's eldest great-niece, a young woman exactly my age named Nyima Putik (figure 3.3). Nyima was one of the young woman traders who spent much of her time outside of the village, selling sweaters in Assam for the months of October through February. I empathized with

Figure 3.3 Portrait of Nyima Putik, Chomo Khandru's great-niece, 1999
Source: Photo by Sara Shneiderman.

Nyima when Chomo Khandru constantly berated her for succumbing to the religion of money; Nyima was just like any other young woman in the 1990s anywhere in the world, seeking new experiences and independence from her family. In many ways, she was just like me.

I began to suspect that Chomo Khandru's harsh attitude toward Nyima's cohort belied a sense of personal responsibility for failing to provide the young women with a better religious education, or to inculcate her successors with a greater sense of spiritual commitment. Nyima Putik was in fact very similar to Chomo Khandru herself—a strong, self-confident woman who cared about her village and family but did not want to be tied down by them and valued the freedom that travel brought. Yet she was not a *chomo*, did not read Tibetan, and was only nominally religious. Nonetheless, she had taken on many of the gender-bending attributes that defined Chomo Khandru's youth: Nyima was the oldest child of six, and since her father had died young and her mother was chronically ill, Nyima became the honorary man of the family, just as her great-aunt had before her. As of 2005, Nyima was 30 and had chosen to remain unmarried, an uncommon state of affairs, especially since her younger sister had already been married for several years.

When I asked Nyima why she did not become a *chomo*, since I thought this might have added some social legitimacy to her otherwise unusual single status, she chuckled. "The *chomo* tradition is not about the future, it's about the past," she said. "You should know that," Nyima continued:

> After all your research, you should know that there is no motivation for us young women to become *chomo*. It's the worst of both worlds: they are not full-fledged monastic practitioners, so no one respects them, but they also don't get to enjoy the pleasures of married life or the freedom and excitement of trade like I do, because they are always worried about if they are doing enough dharma. There is no point in having *chomo* any more—if women want to practice religion they should become proper nuns in one of the big places, and if they want to stay unmarried they can just be businesswomen like me. Why bother to become something that no one cares about?

Nyima's attitude shows that while general social expectations of women have become more liberal—in Chomo Khandru's time it was virtually impossible to remain unmarried without becoming a *chomo*—religious expectations of women have become more conservative. Appreciation for the "practical dharma" that Chomo Khandru exemplified has diminished as the "dharma of study" has taken precedence. Women who could previously gain respect as lay practitioners by taking teachings in their home village and demonstrating their commitment to dharma in action must now leave the village to join a monastic institution elsewhere if they wish to be considered genuine. The problem with this equation is that there are few Bon institutions that offer formal religious education for women in Nepal or India, so Lubra women who want to be nuns must give up their Bonpo identity to join a Buddhist institution.[21] While some women have done this in recent years, many more have decided to eschew religious practice altogether and take up the trading life like Nyima Putik. While most of them travel only as far as India, the more adventurous among them have migrated to the United States, where they work as illegal nannies and manicurists (Craig 2002). At least in these roundabout ways they can maintain their lay Bonpo identity (which they would have to give up in order to become a formal nun), and, by virtue of their economic successes, gain a status level within the community much higher than that of a *chomo*. For all of these reasons, the highly fluid religious identities that women held in Chomo Khandru's time have given way to a more limited set of possibilities that have been shaped by outside actors like the Nepali state, Tibetan exile monastic institutions, and foreign interests. When Chomo Khandru died in 2002, the Bonpo *chomo* tradition of Lubra went with her.

I sometimes wondered whether Nyima Putik and I were two halves of a whole when it came to our relationship with Chomo Khandru. Despite having a somewhat contentious relationship with her great-aunt, Nyima Putik lovingly provided for Chomo Khandru's material needs in her old age, while I gave the old woman a chance to talk about her life, lending legitimacy to her experiences through my interest. Together, by listening to Chomo Khandru's story, continuing the tradition of trade and travel, and honoring her approach to spiritual life, we have continued her lineage—if not as *chomo*, as independent women of the world who believe in practical dharma.

Acknowledgments

I would like to thank Sienna Craig, Charles Ramble, Pat Symonds, Mark Turin, and Mark Unno for their invaluable contributions to my work on this topic over time, and the editors of this volume for their insightful comments on this chapter.

Notes

1. There are two Tibetan etymologies for this word: *chos mo* literally means "woman of the dharma"; while *jo mo* means "noblewoman" or "woman of high rank." Both meanings inform indigenous understandings of the term. Although the term is often translated as "nun," I prefer to use *chomo* throughout, since

"nun" suggests an individual who has taken formal vows within an institutional setting, which is usually not the case for *chomo*.

2. Although my work with Chomo Khandru certainly began as an ethnographic encounter that I initiated in my role as an anthropology student, it took on a more personal, spiritual dimension over time as it became clear that Chomo Khandru began to think of me as the disciple she had never had. While I cannot know exactly what she thought, her actions and words suggested that she experienced our growing relationship as an opportunity to pass on her understanding of the dharma in the hopes that I would find it valuable and share it with others in turn. It is in this sense that I use the term "transmission."
3. Western feminist writers on Buddhism have in various ways attempted to revalorize Tibetan women's history as an inspiration for contemporary Western female Buddhist practitioners, while at the same time distancing themselves from modern Tibetan and Himalayan women's position within patriarchal cultural and religious contexts (Campbell 1996; Shaw 1994; Klein 1994; Gross 1993). These approaches do not adequately acknowledge contemporary indigenous women as full subjects with their own forms of spiritual agency. See Shneiderman (1999) for the details of this argument.
4. Academic scholarship on Tibetan religion must be considered as another external entity that has affected women's religious identities in Himalayan areas, but a thorough discussion of this issue is beyond the scope of the present article.
5. Although we might assume that this opposition is also one between the "modern" dharma of study and the "traditional" dharma of action, such a reading is too simplistic. The dharma of study always existed in Lubra—long before the notion of "modernity" did—as an important form of religious practice, alongside the dharma of action. In earlier eras these various forms of religious identity were for the most part seen as equally valuable, while in recent years the dharma of study has eclipsed the others in perceived prestige.
6. Other authors have commented on these dynamics. See Sihlé (2002) for a general discussion of the increasing monasticization of ritual practices in the Mustang area. In an insightful analysis of "culture change in the name of cultural preservation" in Nubri (an ethnically Tibetan area in western Nepal's Gorkha district, just east of Mustang), Childs states that, "An unintended consequence of foreign patronage for Buddhist monasteries in exile has been a loss of Tibetan cultural diversity in Himalayan highland communities" (2004).
7. Baragaun means "twelve villages" in Nepali. The Nepali government officials who gave it that name misconstrued the locally defined boundaries of the larger community, excluding seven of the member villages.
8. Earlier scholarly works in English generally used an umlaut over the "o"—"Bön"—to approximate the Tibetan pronunciation of the word, but more recent works have dropped the umlaut. For simplicity's sake I have followed the latter convention.
9. For more detailed information on Bon, see Karmay (1972, 1998); Snellgrove (1980); and Samuel (1993).
10. Bonpo literally means "Bon person" in Tibetan. "-po" is a male nominalizer, and technically when referring to women it would be preferable to use the

female "-mo" or "-ma." However, since "Bonpo" has emerged as the standard term to refer to all Bon practitioners in the scholarly literature, I have followed that convention here.

11. As described by Ramble (1984), Lubra is a priestly village, with each household having its own hereditary lineage. In theory there should always be 14 married householder lamas in the village, one from each house, but many hereditary lamas of the current generation have stopped practicing actively. Those who do maintain their lineage are often away trading or conducting rituals elsewhere, so it is now rare to find more than three to five lamas in the village at any time.
12. Cassinelli and Ekvall offer a description of a similar "nun tax" in the Tibetan principality of Sakya (1969:296–297).
13. See Schuler (1978, 1987) for details of marriage practices in Baragaun and Levine (1988) for additional information on polyandry in the Himalayas.
14. *Gompa* means "any solitary place where meditative practice can be carried on" (Snellgrove 1957:200–201). Although the word is often translated as "monastery," this is a misnomer in a village like Lubra, where the *gompa* has no permanent inhabitants.
15. It is hard to judge the validity of such statements from someone like Chomo Khandru whose character was infused with humility. Her responses to questions about her own achievements were always self-deprecating.
16. See von Fürer-Haimendorf (1975); Fisher (1986); and van Spengen (2000) on the trans-Himalayan trade.
17. Although the salt-grain trade ended with the 1959 Chinese occupation of Tibet, new items and routes of trade have developed over the last several decades. Most contemporary Baragaun men and some women spend three to five months of the year in northeastern India selling sweaters and other manufactured goods.
18. The Tamang are a major ethnic population living in Nepal's middle hills, who speak a Tibeto-Burman language, practice their own form of Buddhism, and maintain strong links to ethnic Tibetan populations elsewhere in the Himalayas.
19. See Goldstein (1997) and Shakya (1999) for details of these historical events.
20. See Moran (2004) for an insightful discussion of the dynamics between Tibetan and Western Buddhists in Kathmandu.
21. There are relatively large Bon nunneries in the eastern Tibetan area of Amdo (within the Qinghai and Sichuan provinces of China), but these are geographically and linguistically inaccessible to Lubra's women. However, the last decade has seen a revival of Sakya and Kagyu Buddhist nunneries in the Baragaun and larger Mustang area, creating many more opportunities for women from these sects to join such institutions than were available in the past.

Bibliography

Campbell, June. 1996. *Traveller in Space*. New York, NY: George Braziller.

Cassinelli, C.W., and Robert Ekvall. 1969. *A Tibetan Principality: The Political System of Sa-skya*. Ithaca, NY: Cornell University Press.

Cech, Krystyna. 1993. The Social and Religious Identity of the Tibetan Bonpos. In *Anthropology of Tibet and the Himalaya.* Charles Ramble and Martin Brauen, eds., pp. 39–48. Zurich: Ethnological Museum of the University of Zurich.

Childs, Geoff. 2004. Culture Change in the Name of Cultural Preservation. *Himalaya: Journal of the Association of Nepal and Himalayan Studies* 24(1–2):47–58.

Craig, Sienna. 2002. Place and Identity Between Mustang, Nepal and New York City. *Studies in Nepali History and Society* 7(2):355–403.

Fisher, James. 1986. *Trans-Himalayan Traders: Economy, Society, and Culture in Northwest Nepal.* Berkeley: University of California Press.

von Fürer-Haimendorf, Christoph. 1975. *Himalayan Traders: Life in Highland Nepal.* London: John Murray.

Goldstein, Melvyn C. 1997. *The Snow Lion and the Dragon: China, Tibet, and the Dalai Lama.* Berkeley: University of California Press.

Gross, Rita. 1993. *Buddhism After Patriarchy: A Feminist History, Analysis, and Reconstruction of Buddhism.* Albany: SUNY Press.

Gutschow, Kim. 2004. *Being a Buddhist Nun: The Struggle for Enlightenment in the Himalayas.* Cambridge, MA: Harvard University Press.

Hobsbawm, Eric, and Terence Ranger, eds., 1983. *The Invention of Tradition.* Cambridge: Cambridge University Press.

Karmay, Samten. 1972. *The Treasury of Good Sayings: A Tibetan History of Bon.* London: Oxford University Press.

———. 1998. *The Arrow and the Spindle: Studies in History, Myth, and Ritual in Tibet.* Kathmandu, Nepal: Mandala Book Point.

Khandelwal, Meena. 2004. *Women in Ochre Robes: Gendering Hindu Renunciation.* Albany: State University of New York Press.

Klein, Anne C. 1994. *Meeting the Great Bliss Queen: Buddhists, Feminists, and the Art of the Self.* Boston: Beacon Press.

Levine, Nancy. 1988. *The Dynamics of Polyandry: Kinship, Domesticity, and Population on the Tibetan Border.* Chicago: University of Chicago.

March, Kathryn. 1984. Weaving, Writing, and Gender. *Man* 18:729–744.

Moran, Peter. 2004. *Buddhism Observed: Travelers, Exiles and Tibetan Dharma in Kathmandu.* London: RoutledgeCurzon.

Ramble, Charles. 1983. The Founding of a Tibetan Village: The Popular Transformation of History. *Kailash: Journal of Himalayan Studies* 10(3–4): 267–290.

———. 1984. The Lamas of Lubra: Tibetan Bonpo Householder Priests in Western Nepal. D. Phil. Thesis, Hertford College, University of Oxford.

Samuel, Geoffrey. 1993. *Civilized Shamans: Buddhism in Tibetan Societies.* Washington, DC: Smithsonian Institution.

Schuler, Sidney. 1978. Notes on Marriage and the Status of Women in Baragaon. *Kailash: Journal of Himalayan Studies* 6(2):141–52.

———. 1987. *The Other Side of Polyandry: Property, Stratification, and Nonmarriage in the Nepal Himalayas.* Boulder, CO: Westview Press.

Shakya, Tsering. 1999. *Dragon in the Land of Snows: A History of Modern Tibet Since 1947.* New York: Columbia University Press.

Shaw, Miranda. 1994. *Passionate Enlightenment: Women in Tantric Buddhism*. Princeton, NJ: Princeton University Press.

Shneiderman, Sara. 1999. Appropriate Treasure: Reflections on Women, Buddhism, and Cross-Cultural Exchange. In *Buddhist Women Across Cultures*. Karma Lekshe Tsomo, ed., pp. 221–238. Albany: State University of New York Press.

Sihlé, Nicolas. 2002. *Lhachö [Lha Mchod]* and *Hrinän [Sri Gnon]*: The Structure and Diachrony of a Pair of Rituals (Baragaon, Northern Nepal). In *Religion and Secular Culture in Tibet: Tibetan Studies II*. Henk Blezer, ed., pp. 185–206. Leiden: Brill.

Skorupski, Tedeusz. 1981. Tibetan g-Yung-Drung Monastery at Dolanji. *Kailash: Journal of Himalayan Studies* 8(1–2):25–43.

Snellgrove, David. 1957. *Buddhist Himalaya*. London: Bruno Cassirer.

———. 1980. *The Nine Ways of Bon*. Boulder, CO: Prajna Press.

van Spengen, Wim. 2000. *Tibetan Border Worlds: A Geohistorical Analysis of Trade and Traders*. London: Kegan Paul.

Watkins, Joanne C. 1996. *Spirited Women: Gender, Religion, and Cultural Identity in the Nepal Himalayas*. New York: Columbia University Press.

Willis, Janice D. 1989. Tibetan Ani-s: The Nun's Life in Tibet. In *Feminine Ground: Essays on Women and Tibet*. Janice D. Willis, ed., pp. 96–117. Ithaca, NY: Snow Lion.

Chapter 4

The True River Ganges: Tara's Begging Practices

Kristin Hanssen

"It's beautiful. It's red," Tara said in a letter to me describing the cow she bought with the money I had sent. Her mention of the red cow recalled to me the times when she had spoken of the traits of this specific color. Red was the color her guru had given her when she became initiated as a Vaishnava Baul. It reflected the power of menstrual blood containing female procreative seed, esteemed by Tara and her kin for its beneficent properties. Other bodily secretions such as urine, tears, and mother's milk were also held to be benevolent, yet menstrual blood was particularly valuable. Tara described the female substance as the true river Ganges, bearing the same sacred powers in concentrated form. During her ceremony of initiation, when she received a *sannyas* mantra, her sari, blouse, and underskirt had all been red. The color varied from pink and crimson to bright brown and saffron and, like other Bauls, Tara wore these shades in combination.

In this chapter I examine two seemingly divergent spheres, showing how ideas about menstrual blood and related substances enter into Baul begging practices. Drawing on conversations with Tara, a married female Baul in her early thirties, I argue that Bauls attempt to forge social ties to people by channeling their life force, arising from their seed, to others when they sing. In doing so they challenge Brahmin claims of ritual superiority. Moreover, making the step to become a renouncer has, as Knight (this volume) points out, the added implication of enabling female Bauls to leave the household sphere by taking up the practice of singing songs for alms. Significantly too for Tara, becoming Baul allows her to embrace an alternative interpretation of womanhood, one that celebrates a woman's flow as pure, thus inverting Brahmin rules of purity. From Tara and her husband's point of view, singing for alms is a means of making people who are strangers at the outset into one's own. While the songs they sing bear the capacity to nourish others, the alms they receive in return are viewed as tokens of devotion and appreciation. Within this framework, connecting with others is important. Becoming Baul, then, does not involve detachment from society. Tara and her husband continue to cultivate their ties to neighbors, friends, and family, and they also strive to enlarge their social network through practices of begging. Yet, taking *sannyas* is still regarded as a sorrowful event insofar as it implies making the vow not to bear children and committing oneself to a life of hardship by having to rely on alms. While a number of Bauls do in fact have children, the vow made during initiation is regarded as a serious pledge that should be made in earnest. Moreover, the image of the archetypical renouncer, leaving his family behind to wander among strangers, has a powerful hold on people's imagination. Speaking of her own initiation that ocurred when she was 15, Tara said, "People watching me take *sannyas* cried during the ceremony. They knew that I would stay at home with my parents and younger brother, but seeing me dressed in red made them *think* that I would leave, and this is why they cried."

The following chapter is based on ethnographic research carried out between 1995 and 1997 in Chilluri, a village located in Birbhum district, West Bengal.[1] During the initial months I stayed with an older couple, but later I moved in with their daughter Tara, who lived in a one-room adobe house across the road with her husband, their 12-year-old son, and her husband's unmarried sister. Throughout my stay, I sang with Tara and her family whenever they performed at religious gatherings, and at times I accompanied them when they went singing on the trains.

Chilluri is situated at a two-hour's journey on foot from the Tantric shrine in Tarapith, a pilgrimage center where people come to seek the blessings of the Goddess Tara and to cremate their dead. Bauls and Tantrics lie buried near the burning ground, albeit in different sections to mark their separate orientations. The orders are distinct but they nevertheless bear certain traits in common. Men as well as women are admitted, and most are recruited from the lower castes. They conceptualize the human body as a microcosm, mirroring the larger world, and uphold the view that the sacred power (*shakti*) of the gods is possessed by human beings, present in menses and other bodily secretions, and that these may be ingested to achieve longevity. In view of these similarities, and since they hail from the Bengali region, it is generally inferred that the Baul tradition, dating from the sixteenth century, has been influenced by Tantrics, an order whose formation may be traced to the fourth century.[2] Yet, obvious distinctions may also be delineated.

Tantrics worship Shiva and Kali in their terrible and ferocious aspects. And while Kali is a goddess who combines erotic and maternal traits, Bauls direct their worship toward the childless couple Radha and Krishna noted for their grace and beauty. Their orientation may be gleaned from their appearance. Whereas Tantrics wear their hair matted and mark their foreheads with vermillion, Bauls oil and comb their hair and mark their foreheads with a white paste made from sandalwood. A further distinction lies in the fact that Bauls, as opposed to Tantrics, are recruited from the Hindu and Muslim communities. Whereas the latter are known as fakirs, Hindu Bauls call themselves Vaishnava.[3]

That Bauls worship the divine in humans is something that they share with Tantrics. Yet, this is the feature that holds most appeal for the upper middle classes, who tend to portray them as saffron-clad mendicants, roaming the countryside, singing of universal love and their search for the divine in human beings (see Knight, this volume). The entrenchment of this stereotype is largely due to the influence of the poet and nationalist reformer Rabindranath Tagore who, when publishing their songs in the 1880s, interpreted their lyrics in a humanistic fashion. From Tagore's point of view, phrases like "the Man of the Heart" and "Golden Man," so common in Baul songs, signify "the supreme truth of all existence" (cited in Capwell 1986:24). Although aware of the fact that Bauls regard the body as a vehicle for spiritual pursuits in which esoteric rites are central, he believed that these matters should be left to the ethnologist. As a reformer, his concern lay in drawing attention to the beauty of Baul lyrics, and he reproved the tendency of the educated middle classes to seek inspiration from foreign sources, or for attempting to upgrade the Bengali language by employing Sanskrit terms. According to Tagore, Bauls were models to be emulated, able to express universal truths in a simple rustic language without the aid of rituals (27).

The Bauls featured here do not eschew rituals per se, but, as I note below, they do eschew the rituals carried out by Brahmins. Sexo-yogic rites should be performed while young, soon after they are married in the Baul fashion, that is, without the mediation of a Brahmin priest, and the conjugal pair should refrain from bearing children. Indeed, a number of Baul songs allude to the hazardous impact on one's health if a loss of seed is incurred, and a primary aim is therefore to retain it, accomplished through techniques of breathing during coitus by which the seed is drawn back to the head, the site regarded as its principal abode. In contrast to the archetypical renouncer described in Vedic texts, where accumulating seed is a male pursuit, and where renouncers aim to keep the power they derive from semen for their own benefit (Khandelwal 2001:165), Bauls believe that female seed corresponds to the seed in male secretions. Also, the power gained from storing seed is not retained on a permanent basis, but is transmitted to others through the medium of music during their begging rounds. While the two forms of seed bear traits in common, female seed differs in its inherent capacity to change consistency and color during menstruation. Also, women possess more seed than men do, and a portion of the female seed that is manifest during a woman's period may be reabsorbed by applying the fluid to one's forehead or by orally ingesting it. However, neither Tara nor her husband did this on a regular basis explaining that unless proper dietary measures are taken and sufficient rest has been ensured, the female substance may be harmful, due to its intrinsic heating properties (figure 4.1). To benefit from female seed, it is necessary to attain a strong and healthy constitution.

Figure 4.1 Tara and her husband Karun, dressed in their finest, 1997

Source: Photo by Kristin Hanssen.

During conversations, Tara and her husband emphasized that singing is mentally and physically debilitating, and that the ensuing weakness that sets in stems from having transmitted their vitality, arising from their seed, to others during the course of singing. While people listening to their songs are strengthened, Bauls are weakened in the process, and to regain their equilibrium they must maintain a balanced diet consisting of various foodstuffs classified as relatively hot or cold. How to procure sufficient food by means of singing was a constant concern, an endeavor they conceptualized as a form of self-sacrifice, and one they endured for the sake of others.

Conceiving the person as open and unbounded is not peculiar to Bauls, but it is a commonly held view in India. Contact with another, as when accepting a meal or a cup of tea, or when touching the same object entails that one's own nature is transformed. While the properties of other people permeate one's self through social interaction, the transmission of substances include entities that are intangible, like sound or smell, so that listening to a song or smelling the perfume worn by another affects the nature of one's own material makeup (Lamb 2000:32). McKim Marriott, who initially shed light on the Indian person's permeability, suggested that because people do not view themselves as clearly demarcated and self-contained, but dynamic and divisible, bearing an intrinsic capacity to mix with others, Indians are more aptly described as "dividuals," in contrast to Europeans and Americans who see themselves as individuals, that is, indivisible, stable persons (1976:111). The theory offered by Marriott cogently sheds light on the way in which the subjects I describe actively and consciously made an effort to regulate their flow of seed according to their diet and conditions of the weather. Moreover, by absorbing other people's seed, such as the seed inherent in their guru's food leavings and his urine, they also absorbed the benevolent traits embedded in this substance. As a result, their mental and physical constitution was enhanced. Similar perceptions are held by nonrenouncer villagers, which is evident by the widespread preoccupation as to what to accept and what to decline when offered food by others. As Trawick notes for Tamils, people worry about whom one should associate with and whom one should avoid, especially if the person is untouchable, since mixing with others may entail that you become like them. Through touch, gaze, or otherwise, they will be "part of your system" (1992:99).

While the separation and ranking of caste differences has been a major focus of research on South Asia, I concur with Lamb's observation that processes of mixing are also quite pervasive (2000:35–36). Like the villagers described by Lamb, Tara interacted with kin and neighbors on a daily basis. Tara's parents often came to eat with their daughter's family. Tara's husband's two brothers, neither of them Baul, lived on the same block in separate households with their wives and children, and they as well as other neighbors frequently dropped by to borrow spices, to gossip, or to watch TV. People spoke of being attached to the soil and the water of their village, and like the Tamils discussed by Daniel, they believed that the quality of the soil mixes with the substantial traits of its inhabitants (1984:77–79). As such, Bauls do not differ from their fellow villagers in their striving to retain webs of connection and mutual dependence on neighbors, kin, and friends. Where they do differ, however, is in their striving to attain new ties to strangers hailing from other villages that crosscut caste and class relations. One afternoon, while Tara and I were sitting on the steps leading out of her courtyard, she lowered her voice and, nodding her head in the direction of the neighbor women, seated on the opposite side of the road from us, she said, "Our neighbors do not go anywhere. Except for family relations, they do not meet with other people. Since we sing on the trains, we meet literate people like you who come to visit us and hear us sing."

Singing

The train where Tara went singing was a local train passing through a nearby town, just a 20-minute walk away from Chilluri, beyond the market and across the field. Usually her husband joined her, in which case he would sing while she beat the rhythm with a pair of cymbals. After finishing a song, Tara's task was to open up her patchwork begging bag and move down the aisle, holding it out to passengers so that they could drop a coin or rupee note into it. If her husband was unwell, her sister-in-law Kalpana accompanied her, and the two would sing in unison.

Whenever they returned after singing on the trains, Tara and her husband would enter the courtyard, looking flushed and tired. After splashing water on their feet to wash away the dust, they would seat themselves on the shady porch, accepting the goblet of water that Kalpana hurriedly offered them, while I waved a palm leaf fan to cool them down. Aside from that, my task was to count the money. "Because you are educated," Tara said. The amount varied from fifty to a hundred rupees, most of it in coins, although once I found a ten rupee note, and seeing my look of surprise, Tara's husband said that there was a wedding party on the train and that the groom gave fifteen rupees altogether. "Receiving ten or fifteen rupees makes you happy," he remarked, smiling wanly from the cot where he lay resting.

What worried Tara and her husband most was the effect that singing had upon their constitution. "The life of a Baul is hard," said Tara. "We grow extremely weak." She suffered spells of dizziness when singing, and when walking in the heat of the sun to and from the station. And, since the trains were often late, she did not get home early enough to eat her mid-day meal on time. These factors upset her balance, leading to an accumulation of hot gas in her stomach, rising to her head, and she worried that the gas might cause a stroke. Dizziness signified an oncoming stroke, so she refrained from singing whenever she felt faint, rubbing her body with coconut oil and pouring water on her head, both of which were cooling. Her husband suffered from the opposite condition. To protect himself against the cold thought to stem from moisture, rain, or dew, he always heated up the water he employed for bathing, and sometimes he refrained from bathing altogether. As a further remedy, Tara rubbed his body with mustard oil in which garlic had been fried; both are viewed as hot. But Karun was not entirely immune to the heat. Sometimes he had diarrhea, gas, or a headache, and there were times when he felt hot and cold simultaneously, a condition that several of his village neighbors also suffered from. As Daniel states for Tamils, although people are graded along a continuum of relatively hot and cold, their dispositions are not stable. Depending on the circumstances, a person's thermodynamic qualities may shift in one direction or the other. Correcting the imbalance is done by means of food (1984:186; Alter 1992:120). I once asked Tara whether their strength might be restored by ingesting menstrual blood, a substance bearing nourishing and heating properties. But she said that doing so would be like taking poison, and pointed out that the change from cold to hot or vice versa must be gradual and that exorbitant temperatures of either kind must be avoided.

Mulling over their predicament, I brought the matter up again a few days later, asking Tara why singing had such a debilitating effect on them. She paused in her act of cutting vegetables. Looking up at me she said, "A song is a thing that has a life." She measured out a little space between her thumb and index finger, as if holding up a song, saying, "Even though you cannot see it, there is life within a song." "But," she said, "the life expires if a song is not performed correctly. Not everyone understands the nature of a song. When you sing, the melody must soar. If low it will be ruined." A good singer is capable of singing in a sharp, piercing manner. "Such a voice is sweet," she said. "Because a woman's voice is sweeter than a man's, most people favor songs performed by women." I told her I had played a song of her husband's that I recorded for a friend of mine in Norway who thought the singer was a woman. At this she smilingly exclaimed, "Yes, my husband's voice is good."

Tara's father gave a similar description. When I queried him, he grabbed my pen and made a tiny zigzag pattern in my notebook, stating that the melody should sound like little drops of dew, and to exemplify, he started singing. As he did, he let the melody meander in a graded fashion, ascending along a scale, when suddenly he stopped. He proceeded to explain that a singer must sing rhythmically. By comparison, he said, their neighbor Jaggadish, another Baul, sounded like a railway train when singing in his customary low and rapid fashion. When Tara's father became bedridden with tuberculosis, Jaggadish assumed the role of accompanying me during our practice sessions as well as during concerts. He had learned to play the *dotara* (a lute-like instrument) from Tara's father, which may explain the criticism he frequently received, for Muni Baba repeatedly told him that he played too fast, and that in doing so, he forced me to sing in the same rapid fashion. As Alter notes for Banaras wrestlers, within a Hindu framework, breath control is held to be the cornerstone of any exercise (1992:96, 107). For Bauls, controlling one's breath is the principal means of controlling one's flow of seed, from which the life force (*pran*) arises and which is directed into the melodies. When I brought the matter up with Tara later in the day, she described their neighbor's throat as "fat," explaining that a song loses its capacity to nourish if performed quickly and monotonously. For a song to come alive, a singer must breathe life into the song. Yet, while those who listen are empowered, the singer is enfeebled, and the ensuing weakness that sets in renders her or him susceptible to weather conditions such as the heat of the sun or the cold from the rain, leading to a loss of equilibrium. Tara said that the onset of her father's illness resulted from his lifelong practice of singing songs for alms. Because his strength had been diminished, he had caught tuberculosis, believed to be a cold-related sickness.

Tara's Road to Renunciation

Because Tara's younger brother lacked talent, he had never learned to sing. Instead their father had focused his attention on Tara. "I was seven or eight years old at the time, too young to wear a sari," Tara said when speaking of this period of her life. Learning to sing had been a trying experience. She recalled that her father would slap her if she failed to sing correctly, and that her mother had tried to intervene by telling him to let their daughter be, to which her father had retorted, "Singing means food." Tara, however, was quick to point out that her father had also encouraged her by poking fun at her for feeling bashful about singing on the trains, drawing her out by asking her good humoredly, "Are you feeling shy?" Not only did singing on the trains mark her as different, kids her age would tease her because she wore a pair of rosaries—strings of sandalwood beads tied around her neck. Associated with Vaishnava renouncers, rosaries are also worn by older men and women who devote more time to religious worship as they advance in age. In the eyes of her peers, a young girl wearing rosaries may have seemed incongruent insofar as they signify foregoing the pleasures of the world like the prospect of a lavish wedding, a large and growing family, and accumulating riches. Renunciation entails a simple life, and although some consider such an undertaking admirable, others view it as pretentious, which is why many people hesitate to wear them. Smiling, Tara said, "I'm not embarrassed. I'm Muni Baba's daughter," implying that she resembled her father in having no qualms about disregarding social conventions by taking up the life of a Baul.

Tara was six or seven when her father became Baul, but neither she nor her mother knew the details of the circumstances leading him to make this step. "I don't know. You must ask my father," she said, when I questioned her initially. Her father, however, evaded the issue, though he did attempt to provide an answer. Evoking an image of past grandeur he said, "My father, grandfather, and great-grandfather were all Bauls." Yet his wife's remarks suggested otherwise for she described the way her husband used to look during the early phases of their marriage. He had made a living as an actor impersonating women, wearing a nose ring, a sari, and two coconut shells fastened to his chest, made to look like breasts. Giggling, she told me that she used to feel embarrassed when watching him perform, hiding her face in the free end of her sari. "Then what happened?" I inquired. Her mood changed abruptly, and angrily she yelled, "I had to sleep over there," nodding her head toward a shabby room beside the hearth. "He never gave me food. He left." Then she began to cry.

In the weeks that followed, Tara's father was taken seriously ill. Seeing him grow weaker and accompanying me to the pharmacy and to the doctor's office to obtain medication and advice as to how we should care for him undoubtedly gave Tara cause to reflect upon the way her father had formerly behaved toward them. One day as we were walking to the market, she began to speak about her father's past. She confirmed what her mother had previously told me: that he had abandoned them at a time when she and her brother had been much too small to take care of themselves, leaving their mother with the sole responsibility. "This is why she gets angry at my father," she said in an attempt to shed light on the fact that her parents frequently argued.

On another morning as I was sipping tea on their doorstep, watching the women fetch water from the tube well across the road, Tara sat down next to me. She said that when her father left he went to Nawadvip. "Oh," I said, "Nawadvip," but Tara suddenly hesitated: "You must ask my father, I'm not quite sure." She went on to say that her father had been gone two or three years and that during his absence, he had been learning how to sing. When he eventually showed up at her mother's doorstep, he was dressed in saffron, wearing a full beard, with a *dotara* slung across his shoulder. By his side was a woman clad in white, a stranger to them, who greeted Tara's mother, saying, "This is your husband." Tara believed this woman, whom she and her mother fondly nicknamed Bara Ma (great mother), had been her father's guru, that she had taught him Baul songs and how to worship Krishna. It was due to her efforts that her parents had been reunited. Bara Ma remained with them for a year or so, during which people in the neighborhood grew increasingly attached to her. At dawn, she used to walk around the village, singing songs outside the neighbors' homes, songs beautiful to wake to. Tara said the reason why she left was because her parents constantly argued. One day, when Tara's father struck his wife severely, causing her to bleed, Bara Ma exclaimed, "What are you doing? She is your goddess!" Unable to endure their fighting, she collected her belongings and moved away. It was after Bara Ma's departure that Tara started singing Baul songs with her father. The relationship between her parents continued to be strained, however, and deteriorated greatly when her father took another wife. His second wife, also a Vaishnava sadhu (ascetic), gave birth to a baby girl named Dipa. The three of them slept in the main room, while Tara, her mother, and brother were obliged to sleep in the little storage room beside the hearth. Tara said that their house was never peaceful and that this was why her father left a second time. Eventually he did return, and though Tara and her parents no longer kept in touch with Dipa's mother, Dipa herself, now a grown woman with children of her own, came to visit now and then, and she gave news of her mother's affairs and on how the two of them got by begging with a group of other women.

The House and the *Yoni*

Her father's conduct notwithstanding, Tara was proud of being Baul and of the knowledge she had gained about the properties of seed, why it should be retained and periodically ingested. Speaking of female seed, she said that it bears all the rudiments of a person: not only bones, flesh, and hair, but also thoughts and feelings. Feelings and thoughts embedded in seed travel around the body though various floral centers, but return to the head where the substance is envisioned as a pond, from which the true river Ganges flows. Although the petals of the flowers are usually closed, they unfold each month for a three-day period, and this is when the seed in menses emerges from the female organ. The songs that Tara and her husband valued most spoke of rivers overflowing and flowers blooming, alluding to a menstruating woman. For me the songs were a principal means of gaining access to their knowledge that should be concealed and not revealed to householders, since in ideal terms it is reserved for and transmitted orally by sadhus.

One day while we were by ourselves, I asked Tara and her husband to explain a song that I had learned from Tara's father, composed by the nineteenth century singer Lalan Fakir.[4] A line in the song had puzzled me, which spoke of thefts committed in the body, so I asked them what the thefts might signify. In answer to my question, Karun got up and stood on the porch, while Tara explained the setting. "My husband is a man, while our house is a woman. He is a penis [*linga*] while our house is a vagina [*yoni*]." As she finished speaking, Karun crossed the threshold and entered the house. Once inside, he turned about, gazing at the light blue walls, yellow in the light from the kerosene lamp, and called, "Now I'm inside, and I'm looking around. Everything is very nice," after which he stepped back onto the shady porch. I watched his orange-clad figure pausing in the shade. Then slowly, he turned and walked into the house again, calling, "Now I'm back inside. But this time, I'll commit a theft." His eyes swept across the lamp-lit room. Seeing my thermos, he grabbed it and carried it to the porch outside. The demonstration was over and the question posed was directed at me. "Which is better?" said Karun. "Enjoying the beauty of the room, then leaving? Or, entering the room and stealing something?" I said, "Viewing the room, but leaving it untouched is preferable," at which point Tara exclaimed, "Yes, because getting the thermos is like getting a child." To make sure that I would understand, she added, "Entering a house to obtain a thermos is like entering a woman to obtain a baby. You rob the place of valuables." Nothing more was said. Yet, the message she conveyed was that seed should not be transported out of a woman's body in the form of a child. For although part of their vitality is transmitted to others when they sing, the greatest loss is suffered if one conceives a child.

When speaking about female seed, Tara never opposed it to male seed but simply said that for a woman to conceive, the two must merge. When they do, part of their love is lost. "Their love goes to their baby," she said while pointing to the floor as if to an imaginary child, thus indicating that having children entails that the pair will cease to love each other. She said, "People must learn the proper mode of worship at an early age, so that they do not lose their seed." Karun added, "Lots of people choose to become renouncers when they grow old. At first, they marry, work, and raise a family, and when their children have married and settled down, they become renouncers, but by that time it is too late, for their seed is already spent." Thus the fourth stage in the life of a householder, described in Vedic texts as the period when a man should leave his family to subsist on alms, is not a viable alternative since to sire children means that you become emotionally, mentally, and physically depleted. To retain one's seed is not enough however. What a person eats has direct bearings on the quality of seed (cf. Alter 1997:283–284). An excess consumption of heating products (meat and eggs) leads to a build up of abominable emotions like anger, greed, desire, and envy.

During conversations Tara explained that seed from which thoughts and feeling spring is found in all the substances emanating from the body, including tears, sweat, mother's milk, feces, urine, and saliva. Of these, urine and saliva are the fluids most commonly transmitted by a guru to his or her disciples. While saliva is transmitted through leftover food, gurus will transmit their urine by allowing disciples to absorb it orally or by smearing it onto their skin. I asked her whether feces should also be ingested, to which Tara answered yes, explaining that because it improves the quality of the vegetation, humans should fertilize their bodies by absorbing a part of the feces they emit. "There is life in feces," she said. "It makes you healthy." As Tara spoke she had been busy peeling potatoes, dousing them in water. Suddenly she paused. Looking up at me, her face assumed a worried expression as though she wondered whether I would disapprove. "This is right talk, isn't it?" she queried. She went on to explain that seed emerges from nine openings in the body, then proceeded to recount them as consisting of the eyes, the nostrils, the ears, the mouth, the genitals, and anus, adding that seed also emerges through the pores of the skin in the form of perspiration, though the goal is to prevent this from occurring. "Great sadhus can sweat, defecate, and urinate without emitting seed," she said.

A few days later, Tara suddenly exclaimed, "You are not a sadhu if you do not have a spouse. A lone person's seed goes bad when it trickles out of the body." Referring to her unmarried sister-in-law she said, "Kalpana is not a sadhu. When she sleeps she dreams and as a result her seed flows out. Letting it fall is a sin. But if you practice *sadhana* [method of worship] with a spouse, the seed stays good," by which she meant that since the nature of seed is to circulate, it must be roused periodically. Once the flow has been instigated its loss is curtailed when raised to the head through breath control.

Nocturnal emissions are regarded as a problem among subjects seeking to retain their seed (Alter 1997: 287–288; Bottéro 1991:307). But within a patriarchal Brahmin framework, efforts to control one's flow is a male pursuit. Indeed, there is little evidence suggesting that women store their seed or that they are concerned about its loss. Instead, women are characterized by superabundance, akin to overflowing vessels, often depicted as temptresses ready to devour men's supply, their source of strength and vigor (Khandelwal 2001:166). However, for Tara, women, like men, are prone to lose their seed while dreaming, and though she may have acquired this perspective from her male guru, there was nothing that she said that conveyed discomfort with this view. Rather, the seed in male and female secretions were said to be the same. The difference lay in the fact that women also have menstrual blood, something that men lack, and so men must procure this substance from a woman.

The First Female Flow

We were busy cleaning spinach on the porch when Tara suggested that I fetch my notebook. So I seated myself beside her with pen and book in hand. Since what she was about to say was secret, she closed the door leading out to the road, so that neighbors would not enter. Her face assumed a serious expression, and lowering her voice she told me she was 12 years old when her first flow appeared.[5] Her father gave her a white strip of cloth, instructing her to wear it to soak up the menarche. She did as she was told, after which he delivered the cloth to the family guru. Then one month later, Tara, her parents, and three other sadhus were summoned to their guru's ashram. After pouring water on the cloth, for it had hardened during the intervening period, their guru placed the cloth in a coconut vessel to which various liquids had been added, including cow's milk, coconut milk, and palm juice, ingredients serving to cool the heating properties of Tara's first flow. Sugar and camphor were also added, and the resulting mixture acquired a pink, sweet, and thick consistency. Tara smiled and cupped her hands to illustrate the amount that each of them ingested, saying, "We drank a tiny little bit," implying that because the substance is so powerful a small amount will suffice to benefit from it. She added, "If my son had been a girl, great sadhus would come to my house asking me if they could have a portion of my daughter's flow. Very few people manage to obtain it." Tara said that Radha was the first woman who had offered a portion of her flow to her male partner Krishna. In fact she often referred to Radha as a model, stating that though Radha had been married to a man called Aian she did not lose her seed by having children, for Aian lacked a penis. Instead Radha performed sexo-yogic rites with Krishna and in doing so her seed remained intact. "This," said Tara, "is the reason why people do not call her mother. Exempting Radha, all the other goddesses are termed mothers." Her husband, who was seated beside us frying fish, remarked that I was fortunate to learn about a woman's first flow. "You cannot read about it in a book. This is very good talk," he said, implicitly critiquing the Brahmin custom of employing Vedic scriptures during worship. Tara told me never to disclose what she divulged to householders. "If they knew what we know, they would be disgusted. They would not understand why the seed should be ingested. They may listen to the songs, and they may enjoy the beauty of the lyrics, but we must not tell them what the lyrics signify." At other times, however, she and her husband would suddenly exclaim, "If they knew what we know, they would stop leading the life of householders and become renouncers. This is not a fairy tale. This is true."

Initiation

Tara was initiated into *sannyas* at the age of 15. When I questioned her explicitly about why she chose to become a Baul, she said that Bara Ma had taught her that being Baul is a good thing. Her father was not mentioned as a motivating force though he played a key role in organizing the event, providing flowers, sweets, and incense needed for the rite. Since her mother lacked *sannyas*, she also went through with the ceremony. Tara smilingly recounted the proceedings, stating that she was dressed in new saffron clothing: a sari, underskirt, and blouse. "All my clothes were red," she said. A white cotton cloth had been tied around her forehead marking the site from where the flow of seed descends. Tara remained secluded in a room for three days subsisting on plain rice and potatoes. When the three-day period was over, her guru entered the room to divulge a *sannyas* mantra through her forehead. Then a woman entered to give her a brass plate and a bowl of sweet rice pudding, items signifying alms. Tara fetched the plate of brass to show me, dusting it off to reveal the inscription "benediction" on its surface. She resumed her narrative, stating that the ceremony was like a birth. "This is the why our clothes were new and it is also why we cut our hair. A newborn baby has no hair. Everything is new." Then she added that she and her mother did not remove their hair completely. "You feel bad seeing a woman without hair, so we just cut off a little bit," she said, pulling a few strands from the nape of her neck to show me what was cut. Tara went on to explain that the mantra that her guru gave her contained seed and that by divulging it he became a father to her (cf. Gold, D. 1999:74). The woman who gave alms took on the complementary role of mother, and since it is the woman who introduces a novice to a life of begging, she is commonly referred to as "begging mother."

Tara compared her rite of initiation to the rite a Brahmin boy goes through when receiving a sacred thread, stating that the two ceremonies parallel each other. She brought the issue up while we were visiting a Brahmin family. A man named Indrajit, interested in learning Baul songs, had made it a habit of stopping by on his way from work. During one of his visits, he invited us over for a meal. As we waited to be served, Tara and I walked into the hut where she noticed a picture of him hanging on the wall. It had been taken shortly after he received his sacred thread and showed him wearing a saffron *dhoti* (waistcloth worn by men). Draped around his shoulders was a saffron cloth that partially concealed the sacred thread that ran diagonally across his chest. His head was bald and a pink strip of cotton cloth was wound around his forehead. Tara's face lit up as she excitedly pointed to the picture, stating that this was how her husband looked when he received his *sannyas* mantra. Of course, being a woman, she had worn a sari and had not shaved off all her hair, but the color she wore and her other accessories had been similar. When we had taken leave of Indrajit, Tara continued to speak of the rites as we were walking home, telling me that the two ceremonies were almost identical. The difference, she said, lay in the fact that Brahmins tie their *dhoti* in the customary fashion, by pulling it up between their legs, while Vaishnavas let it hang down below their knees so that it almost reaches their ankles. Also, Brahmins have a pink piece of clothing wound around their heads, while the headband that Vaishnavas wear is white. And, while Brahmins get a sacred thread wound around their upper body, Vaishnavas get a loincloth.

At one point, I overheard Tara's husband evoking this last analogy to the mother of a Brahmin friend of theirs who often accompanied them on *tablas* (drums) during concerts. "The two ceremonies are the same. The only difference is that Brahmins have a sacred thread, while Vaishnavas have a loin cloth," he said pulling at the strap tied around his waist. She refrained from commenting, but her stern expression indicated that she did not approve of the analogy. Nor did her son admit to the likeness. He simply recounted the different steps that closely resembled the ceremony that Tara had been through. His age had been an odd number, for he had been 11 while Tara had been 15. He wore saffron clothing, his hair was cut, he had a headband, a staff, a kettle of water, and he sat in a room for three days eating plain boiled food, after which a married couple entered the room to give him alms. He added, however, that during the three-day period, a Brahmin novice should not see a woman's face, since this would entail becoming a sadhu. Thus, what Brahmins seek to avoid is the principal aim of the Vaishnava ceremony.

Caste

During the months I spent in the village, Tara, Karun, and his sister Kalpana repeatedly told me that Brahmins made the caste system and forced people from other castes to carry out various chores. They decided who would have to make their living catching fish and who would have to make pots and pans. Those belonging to the lower castes were forced to work, farming the land owned by Brahmins. Sometimes Brahmins gave the low-caste men alcohol to drink, after which they raped their daughters, or they did not bother to give them alcohol but simply said, "Give me your daughter." Karun said that Brahmins decided how to worship deities. They made up rules about weddings and funerals, and they determined how long the different castes must fast whenever a death occurs in a family. Whereas Brahmins only fast for a short period, other castes must fast for several days, and while all other castes must eat their wedding meal at night, Brahmins eat their meal during the day. "Brahmins," said Karun, "devised this system in order to make others suffer. Poor people must go without food, but Brahmins make sure that their bellies are full."

During one such discussion I impulsively inquired whether there was a hierarchical difference between the Brahmin subcastes. A Brahmin friend of theirs' had told me that the reason why the Mukherjees rank lower than the Bannerjees and Chatterjees is because they once ate cow's flesh. Jaggadish, who was visiting, said he would ask Indrajit which caste was greater, that because he was a Brahmin he would know. On the following day, as soon as Indrajit entered the courtyard, he told us that the man called Mukherjee had never eaten cow's flesh. Although he wanted to eat the cow's flesh, he did not actually do so. There was a sadhu present and he hid the piece of flesh away, threatening to have the Brahmin circumcised if he did not refrain. Indrajit concluded, "Hindus do not eat beef. Only Muslims eat cow's flesh." No one said anything, so I raised the question again as to which caste was the greatest, but Indrajit wavered in his response. "Before, each caste was separate," he said. "It was possible to tell which caste ranked higher, but nowadays, everything is *khechuri*" (a special kind of dish where lentils, rice, and vegetables are mixed together). He then proceeded to recount the following story.

"Once," he said, "there were four Brahmin brothers whose names were Bandopadhyay, Mukhyopadhyay, Cattopadhyay, and Gangopadhay. The four brothers came to this country, where they settled down and married four women belonging to the Shudra caste. That is why women cannot say, *om*," said Indrajit. "Women are not Brahmins. They do not know mantras." There was a pause. Then Jaggadish turned to me and said, "In this country, women do not utter their husband's name. You Kristin, whenever you speak of your husband, you call him Christopher. Indian women do not do this ever. They say 'My husband' or 'My child's father.' They treat their husbands as if they were gods. They show respect." I refrained from commenting but nodded my head to show him that I understood, when Tara suddenly remarked, "Brahmins think that they are clean, but a lot of them are dirty." A short silence followed. It was difficult to ascertain whether Indrajit had been offended for he remained composed. Karun then told him that a

Muslim friend of theirs abstained from eating onion and garlic as well as meat. At this, Jaggadish began to talk about the meal he would prepare, showing us the vegetables that he had purchased at the market, which put an end to the conversation about the different castes. After they had left the courtyard, Tara said, "I think Indrajit grew angry when I told him that a lot of Brahmins are dirty, and that that is why he left so soon." She sounded slightly anxious, but grew confident while speaking, saying, "Jaggadish and Bara Ma believe that Brahmins are the best caste, that they go directly to god as soon as they die. They do not have to be reborn, the way other people do. But only illiterate people say such things. There is a story in one of my son's schoolbooks in which it is written that Brahmins formerly exploited low-caste people." Tara then repeated what I had heard from Karun: that in the past Brahmins would not let other people read the Bhagavad Gita. Only Brahmins were allowed to read and carry out *pujas* (devotional ritual in Hinduism) since they had learned the appropriate mantras. Tara added, "They do not know the meaning of the mantras that they utter. They say '*ham kling om*' while throwing flowers on the figures of the deities, but they do not understand what they are saying," by which she meant that the knowledge that Brahmins have is inferior to that possessed by Bauls.

Tara's Caste

Despite the fact that caste was a common topic, especially their views on Brahmins, it took me a long time before I learned the name of Tara and her husband's caste identity. Since their contempt for Brahmins was pronounced, and since Tara said that they were looked upon as dirty I suspected that their caste was low, but my efforts to obtain more detailed information were met with responses such as, "Caste is not good. We do not like it," or "I do not know. Muni Baba knows." Muni Baba on his part insisted that his family had been Baul for several generations. At one point I attempted to ask the neighbors, but they responded in similar terms, saying that the caste system is bad and that they disliked it.

Then one summer day in 1997, Tara revealed her caste identity. We had finished eating our mid-day meal and had just retreated into the hut to take a nap in the shady interior, which was slightly cooler than the porch. Tara, Karun, and their son had spread their mats upon the floor and straightened out the quilts, while I was on the bed. Once again I posed the question, though I did not expect an answer, when Tara suddenly replied, "Rai Das." Her voice was barely audible, so I repeated the name loud and clear to make sure I got it right while groping for my notebook. Tara whispered, "Yes," smiling shyly, then she implored me not to tell her father that I knew. "He has forbidden me to tell you."

The incident that led her to disclose her caste took place some minutes earlier. We had finished our meal, and were outside the house, letting our hair dry in the sun when I noticed an older neighbor woman eyeing us strangely and keeping a great distance as she passed us in the road. I asked Tara and Karun why she was avoiding us, and Karun explained that she belonged to the sweet maker's caste (Maira), which was why she felt superior. I said, "Is that a great caste?" to which Karun laughingly replied, "She believes her caste is great." Tara then proceeded to explain, "The old woman thinks her caste is clean and views our caste as dirty, so she tries to avoid our shadow. When her son was younger, the woman used to make him change his clothing when he returned from school. He had to keep his school clothes separate from the ones he wore at home." Tara likened the Rai Das caste to a caste called Muci Das, stating that the two are similar for both are leatherworkers. When I asked her if her relatives had really worked with leather, Tara said that her father's family were cultivators tilling the land owned by Brahmins, but because people believed that their true work was leather, they were still considered dirty. Regarding herself, Tara seemed embarrassed when she talked about her caste. Being a Baul was something she was proud of, but being a Rai Das was something she regarded as demeaning. The subject was therefore rarely broached and I did not press the issue. However, on the following day she related a story about a man whose name was Muci Ram Das. A long time ago, she said, this really happened. "It is not a fairy tale," and she proceeded with her narrative.

There was once a man named Muci Ram Das who worked as a shoemaker, sewing shoes. He noticed a Brahmin who passed him on the road, on his way to the river Ganges to do *puja* (devotional ritual). The man named Muci Ram Das called to the Brahmin and asked him whether he could do *puja* on behalf of him, explaining that he had a lot of work to, which was why he was unable to make the trip himself. The Brahmin complied, accepting the two bananas, which Muci Ram Das handed him. When he got to the river Ganges, he entered it and waded out till the water reached his waist. Then he held out his offerings, consisting of a number of different kinds of fruit: bananas, apples, cucumbers, and various flowers. He carefully placed his offerings in the water. Then he turned back to the shore, but as he emerged from the river, he suddenly remembered that he was supposed to give Ganges the offering from Muci Ram Das too. So the Brahmin turned around, but he did not bother to wade back into the river. He just flung the bananas into the water. Ganges assumed her form as a goddess, and lifted her hands up in the air to catch the gifts from Muci Ram Das. Tara said, "The

Brahmin then realized that Muci Ram Das was a very devoted person. He used to think his own *puja* was better, since his caste was greater, and that Muci Ram Das's worship was insignificant because his caste was low. But after seeing Ganges, he was forced to admit that he was wrong. He therefore sought out Muci Ram Das to tell him what had happened, then told him, 'You are a very devoted person.' " Tara looked at me, smiling shyly as she proceeded to make an implicit parallel between herself and Muci Ram Das, saying, "I do not know a lot of mantras. I say to god: here is some fruit. I am giving it to you. This is how I carry out my *puja*. I think it is a good *puja*. Don't you think so too?" I told her yes. Then, after a slight pause, she proceeded to recount another story concerning Muci Ram Das.

Once, Draupadi invited Krishna to eat at her house. She was going to cook the meal herself and serve it to Krishna. But Krishna said, "There is one person who is an extremely devoted man. His name is Muci Ram Das. I think you should invite him first. You can invite me afterwards." At this point, Tara interrupted her tale to ask me whether I knew who Draupadi was. I said, "Yes, she figures in the Mahabharata." Tara nodded, then continued with her story. When Muci Ram Das received the invitation, he went to Draupadi's house to eat. She placed the food that she had made before him. There were lots of different dishes. There was spinach, lentil soup, curds, and sweets. To make sure that I understood how elaborate the meal was, Tara said, "Really, there were five different kinds of curry dishes." She then proceeded with her story saying that Muci Ram Das began to knead the different curries together, and as he started to eat, the bells began to chime so that the gods would be alerted to his act. But while Draupadi was watching Muci Ram Das eating, she was thinking to herself, "I've made all these different curries, and he's mixing them all together. They are not separate anymore. He will not be able to tell one dish from another." At this, the bells grew silent. Muci Ram Das looked at Draupadi and said: "What were you thinking just now? The bells stopped chiming." Draupadi told him, "I've gone through all this trouble, trying to make each dish unique. Now they all taste the same. How will you understand the different tastes? You've mixed them all together." Tara added, "What he did was strange. It was hard for him to eat." She continued to recount the story, saying that when Muci Ram Das heard the answer made by Draupadi, he told her that he was unable to continue. "I cannot proceed," he said. So Draupadi carried his plate away and started to make a new meal. When she had finished cooking, she brought her various curries out, served the food to Muci Ram Das and seated herself beside him. The bells began to chime as he mixed the curries together and resumed his eating. Tara said that this way of eating was his custom. By kneading the various dishes together, the food changed. It became dirty, and so it was difficult for Muci Ram Das to eat the food. As if to explain his act, she said, "He was a very devoted person," and she concluded by saying that when someone cooks a meal, she should always sit next to her guest, watching the person eat to make sure there is enough food. But she should not say anything or think anything in particular about the manner in which that person eats. "This is the Bengali system," she added matter-of-factly as she got up to attend to her household chores.

Although several things about her narratives had caught my interest, I was especially puzzled by the fact that Muci Ram Das mixed the curries into one undifferentiated mush. Why was this a pious act? And how was it linked to his occupation as a shoemaker? His behavior at any rate, was highly unconventional or, as Tara put it, strange. It occurred to me that his manner of eating was an inversion of the customary rules pertaining to a meal. During my initial stay in the village, Tara had taught me the correct sequence of eating the various dishes given. I should start off with the spinach, she said, then the potato sticks, followed by the lentil soup, then a vegetable dish, and, depending on what was being served, an omelette, a curry, and finally the chutney. But Tara's latter narrative also reminded me of what she had once told me regarding her stay in Calcutta when she was a little girl, during which time the woman who was supposed to look after her had boiled vegetables and lentils together. Food served in this manner was distasteful to her, which was why the experience was traumatic. Muci Ram Das presumably felt similar to Tara, and as opposed to me, he knew how to eat a meal correctly. For when Tara related the story, she pointed out that it was very hard for him to eat the food and in saying this she seemed to be implying that his action was a feat, which in turn entailed that he changed the curries into "dirt" on purpose. Perhaps, I thought, the gods were impressed by his ability to handle dirt. He carried out the deed despite the fact that he disliked it. As such, his action might be linked to his occupation as a shoemaker. For the shoemaker caste he belonged to was certainly viewed as dirty. Although Tara did not draw attention to this fact when she explained the content of the story, she had done so previously, when stating that leather is dirty and that those who worked with leather are therefore also viewed as dirty. I could not, however, induce her to elaborate further on the narratives. To Tara, the underlying message was that people should not think lowly of others. Even though their action may seem inappropriate, inconsiderate, or even downright repulsive, one should nevertheless refrain from thinking badly of them, since what appears to be base may actually be an indication of superiority, purity, and devotion. As such I think it plausible to draw an implicit parallel between the occupation of leatherworkers and Baul practices of ingesting bodily emissions. In identifying these substances with love, in which menses in particular is held to be a manifestation of the goddess Ganges, Tara and other Bauls invert Brahmin rules of purity.

Another important feature is that the gods may have favored Muci Ram Das because he suffered.

Suffering

Writing about women in Tamil Nadu, Trawick notes that worshiping god by means of suffering entails self-sacrifice. It is a pious act—an expression of devotion—inflicted on one's self in a variety of ways: by relinquishing one's sexuality, by self-starvation, or through maltreatment by one's husband. Since women stand in a relationship of servitude to men, they suffer more than men do, and due to their excessive hardships they gain powers known as *shakti*. The accumulation of these powers enables them to heal, which means that others seek them out (Trawick 1991:14–15).

Like the women described by Trawick, Tara and Karun also believed that suffering entailed self-sacrifice. They also pointed out that women are able to suffer more than men are able to. However, they did not explicitly state that suffering entails an accumulation of *shakti*. Still, ideas about suffering were tied to notions of servitude as devotion and when endured they said it might lead to particular boons from god, which in turn would render those who suffer capable of healing others. Self-sacrifice extended beyond issues of the female gender; for example, collecting alms by means of singing was described as a form of suffering too, during which listeners become imbued with the singers' soul or life force known as *pran*. Not only that, but the ability to suffer was a recurrent theme, whenever Tara spoke of village life as it is experienced by low-caste people like herself, or as it had been lived by Muci Ram Das, the protagonist of Tara's tales. As the first story indicated, he bore several traits in common with Tara and her relatives. He was also poor. His caste was similar to theirs. And like them, he had to suffer the humiliations inflicted upon him by people from the Brahmin caste because of his association with leather, which was viewed as dirty. Moreover, like most poor and low-caste people, Muci Ram Das suffered since he worked so hard that he had no time to go to the river to honor the goddess Ganges. Nor was he able to give the goddess expensive fruit as the Brahmin did, for the lowly job of a shoemaker does not pay much. For all these reasons, Muci Ram Das led a difficult life. The goddess Ganges realized this. To show her appreciation she rose from the water to catch the gift of two bananas, something she did not bother doing when receiving the offerings made by the Brahmin, despite the fact that they were lavish. As Gold points out when commenting on a similar story told by leatherworkers who were considered untouchables in Rajastan, this theme is in keeping "with the tradition of low-caste devotion" in which simple-hearted love is valued as superior to the elaborate expensive rites performed by Brahmins (Gold and Gujar 2002:207). In other words, wealth is not important, but the ability to show devotion is. As Trawick states, devotion is conceptually tied to suffering (1991:19).

The second story told by Tara may be viewed in a similar light. Muci Ram Das willingly transformed the various dishes made by Draupadi into a substance resembling dirt, which he then proceeded to eat. According to Tara, this act of inflicting hardship on himself was an act of devotion. To think otherwise as Draupadi had done was wrong, and this was why the bells stopped ringing while he ate the meal she cooked. Tara, however, did not tell me why this should be the case. She simply related it as a fact. Because Muci Ram Das suffered, he was capable of deep, strong love and so the gods favored him, more so than they would a Brahmin. Devotion was his distinguishing trait, but what of the qualities attributed to gods?

Although Tara did not mention this when interpreting the narratives, I suggest that compassion is another aspect that enters into the relationship between poor people and gods. As all-powerful and all-knowing beings, gods realize how hard it is to undergo distress and pain, and so they sympathize with those who tolerate discomfort. Tara touched upon this subject once, while we were traveling home after visiting Bara Ma who resided in a tiny, thin-walled hut of clay. While sitting on the train, she gazed at the villages we passed and mused, "I think god gave us adobe houses to live in so that we would not have to suffer so. For they are cool in the summer and warm up in the winter as opposed to the brick and tiled houses inhabited by rich people, which are hot in the summer and grow cold in the winter." Her remark suggests that the act of suffering does not go unnoticed and that such a person is entitled to a favor, either in this life or the next. When I questioned other villagers as to what would happen to them when they died, most people said that they did not know what lay in store for them. They hoped that they would go to god but were uncertain as to whether this would actually happen. Suffering, however, was said to be a virtue, a good quality, deserving reward, and poor people suffer more than rich people do.

Begging

Singing was a strenuous activity but could also be humiliating, which is probably the reason why Tara and her husband initially ignored me when I asked them if I could come along. The trains were crowded, often late, and some of the passengers appeared to dislike their singing practices. Tara told me that at times when she and her sister-in-law had finished singing, a man or woman would remark, "You seem like well-brought up girls, why do you sing on trains?" implying that an upright woman should remain at home, cooking and caring for her husband and children. People also said, "You don't look as though you're poor. Why do you beg?" to which Tara invariably replied, "This is our dharma [vocation]."

Tara considered begging a duty, explaining that sadhus have to beg to qualify as such. "Even rich and famous Bauls must beg at least three days a year." She elaborated saying that Bauls who cannot sing do their begging rounds by wandering from door to door in nearby towns and villages calling god's name for which they receive rice grains, vegetables, or clothing. Her husband Karun likewise emphasized that renouncers must heed certain obligations. "You cannot dress in ordinary clothes and you cannot cut your hair; you cannot sire children, nor can you till the land. Good office jobs are barred from you. You cannot work for wages. Begging is your only source of livelihood."

During the months I spent in the village, a beggar would stop by once or twice a day, usually in the morning. Most were men and women beyond their childbearing years, dressed in white or saffron, beating a pair of cymbals while invoking god's name. Eventually I started to recognize their faces, but this took time for a month or two might pass before they came around again. Tara said the reason why they came around so seldom is that sadhus should not beg too often in the same village or people might grow wary and cease to donate alms.

Begging, she explained, is a rule-governed activity. Whereas vendors at the market are obliged to donate alms on Fridays, householders do so on Sundays. Hindus however do not give alms on Thursdays since this is the day devoted to worshiping the goddess Lakshmi, the giver of rice grains, and so householders must honor her by keeping what she gives. By contrast, Muslims donate alms on Thursdays, devoted to worshiping the guru, and this is why renouncers, irrespective of their affiliation, seek out Muslim homes on "guru's day."

Tara had never begged in a village, although she sometimes spoke of doing so, especially when returning home from singing on the train. "I'm sick of wearing myself out singing and having to count the money afterwards," she said, adding that receiving food and clothing would be better. Her husband disagreed, pointing out that singing on the trains was much more lucrative. "You can go around in villages begging rice grains the entire day and all that you obtain is three kilos. If you sell the grains to your neighbors, you may earn eighteen rupees altogether, but you can earn as much as twenty rupees singing four songs on trains." Aside from moral and material gains, Karun pointed out that singing on the trains was a means of gaining publicity, increasing their chances of being able to perform at *pujas*, an event they called a "program," which I will translate as "concert." If they could do a concert twice a month or so, they would not have to beg as much since the sum of money that they would receive on such occasions would be greater, thus allowing them to get more rest. Rest was like a juice, they said, serving to lubricate the body so that the seed would flow, ensuring that the body would stay young and healthy.

Not that giving concerts was an easy undertaking. The events took place at night according to a set pattern. The men sang first followed by the women. Because women conventionally sing in a shrill and piercing manner, a woman's voice is considered sweeter than a man's. Thus, whenever we performed, I would sing as soon as the men were through, followed by Tara and her sister-in-law Kalpana. Concerts usually lasted for a period of four hours, but on several such occasions, male members of the audience would get increasingly drunk insisting that we continue to perform for an unreasonable duration. If we were unfortunate, the food served was insufficient, and the payment was accompanied by unpleasant haggling. Yet, there were times when concerts unfolded according to Tara and her husband's expectations: when we were treated as honored guests, paid according to the sum previously agreed upon, and fed a proper meal after we had finished singing. This was how it should be. Still, regardless of the circumstances, opportunities to give concerts were few and far between. To make ends meet, they had to rely on singing on trains.

Tara was distressed by the fact that Bauls do not receive adequate support, and she often talked about former times, when householders supported their local ashram to a greater extent than they do now. She said, "Thirty-five years ago, householders gave renouncers food and clothing. Landowners gave them rice stalks from their fields, so that they could thatch their roofs. Others gave them milk, clothing, and vegetables. Now everything is contrary [*ulta*]." She specified, "The system of giving entailed that when a cow gave birth, people would donate milk to the ashram, where after they would drink what was left themselves," adding by way of explanation that all householders had cows before. She continued, "When the vegetables were harvested, the Vaishnava ashram would first be given a share of the crops, then the householders would eat. People used to believe that lending support to the ashram in their village was a sacred deed."

Aside from lamenting the current situation in which renouncers do not receive adequate support, Tara's description implies the assumption that the lives of renouncers and householders are attuned to one another, that there is an element of reciprocity built into the relationship. Her rendition of the way householders formerly behaved toward Bauls resembles Vallely's description of Shvetambar Jain nuns who are treated respectfully, and where householders perceive the act of giving alms as an act of merit (Vallely, this volume). Tara, then, does not dissociate herself from householders but conceives her interaction with them in terms of a ritual rule-governed idiom. As such, the distinction between the two domains is not absolute, but interrelated parts of a wider social setting (cf. Cort 1999:90, 103).

Conclusion

As I have sought to show, Bauls do not turn their backs upon the world by disconnecting social ties, which is how renouncers are conventionally portrayed (cf. Laidlaw 2000:631; Fuller 1992:18). Rather, they relinquish certain socially constructed roles associated with the domain of householders (Knight, this volume; cf. Cort 1999:103). In becoming a renouncer, Tara made an implicit vow not to bear children, held to guarantee security in old age. She described this step as one bound up with sorrow, making those who viewed her rite of initiation cry. Why this is so makes sense when seeing renunciation as a form of self-sacrifice, for instead of ensuring her own welfare by bearing children, an act which she associates with selfish desire (*kam*), Tara imparts her seed to others through the songs she sings and in doing so she has to sacrifice herself.

In her overview of Tamil women, Trawick notes that women gain power (*shakti*) identified with love through self-sacrifice (1991:17), a statement that recalls Hart's interpretation on ancient Tamil poetry. He writes that though the sacred power attributed to women was potentially malignant, it was also linked to love and that women could accumulate it through chastity, a word which in Tamil translates as "the restraint of all immodest impulses" (Hart 1973:243). As such, chastity does not simply signify fidelity, but the ability to discipline one's self so as to temper malignant forces. Malignancy is bound up with notions about hot and cold. The reason why the two should be kept in balance is in order to temper one's emotions. Doing so, however, requires self-discipline, held to be a form of suffering, a capacity associated with renouncers, members of the lower castes, and women. Writing of the latter, Trawick states that their power and their suffering "are expressed through a greater capacity for feeling, especially feelings of love." She goes on to state that "genuine love, such as is found in religious devotion [*bhakti*] is a power greater than any other." It makes itself manifest through bodily emissions such as tears shed when weeping and the milk her children drink while she is nursing them (1991:19, 20). In a similar vein, the power projected by Bauls is contained in bodily emissions, but it includes the substance sound, which is transmuted into love, bearing the capacity to nourish. As Trawick notes, this form of power has to be transmitted. If not, it disappears (17). In other words, the power gained is there because it can facilitate other living creatures. Like the female renouncer described by Hausner (this volume), Tara employs her religious power to benefit other people. In keeping with Trawick's insights, those fraught with greed, ill will, and desire are by definition rendered powerless since having such emotions shows an inability to control internal heat. This is true insofar as excessive heat, viewed as destructive, turns into love, tranquillity, and happiness when cooled. And attaining a balance between hot and cold is the key to maintaining a healthy, loving, peaceful frame of mind and constitution.

Tara said, "There are many kinds of love." She expanded by noting that love arises between newlyweds, between an older woman and her spouse, between parents and their offspring, between siblings, between friends, between gurus and disciples, and between deities and devotees. Although she did not elaborate, I suggest that what all these forms of love appear to bear in common is that they spring from self-restraint and as such they constitute an offer of one's self. Thus, women are more loving because they undergo behavioral restrictions. They turn anger and frustration into love when submitting to their husbands and their in-laws in the joint family sphere. Similarly, lower-caste men and women endure torment when they handle filthy matter as apparent in the story about Muci Ram Das, who mixed the curries served by Draupadi into an unpalatable substance, an act which Tara termed a gesture of devotion. And sadhus undergo austerities when they resist the desire to ejaculate. In doing so they store the power springing from their seed, after which they cool it so that it grows benevolent, then pass it on to others when they sing. The sound that they emit is regarded as substantial, serving to strengthen others, and if the sacrifice they make is great, they lose their equilibrium. For Tara and others, a loss of balance was an ongoing difficulty, and to restore their equilibrium, food and medication were considered to be paramount. Indeed, a great deal of time was spent discussing food, cooking it, and seeking out doctors if they continued to feel weak. Yet the pain that they endured was not perceived to be in vain for much of what they said indicates that the motivating force leading them to suffer was the reward that they would gain from deities in this life or the next.

Other motivating forces entered in as well. For Tara gender was important for she made a point of stating that Vaishnava Bauls value womanhood by focusing their worship on the female substance. Unlike Brahmins, Bauls allow women to become initiated. Moreover, in making the step Tara gained a sense of autonomy and self-sufficiency (cf. Knight, this volume), since having to beg entails having to leave the household sphere, which in turn enables her to connect with other people residing in other villages.

As a result her social network is expanded. I should also emphasize that to practice the Vaishnava Baul vocation allowed Tara to develop her talent as a singer, which in turn might lead to social recognition. Yet, her coexisting caste identity did not cease to be significant. Tara still risked being treated as an untouchable by those familiar with her past. Moreover, most people are aware of the fact that the majority of Bauls are recruited from the lower castes, and as such they lack the social capital of Brahmins (cf. Gutchow, this volume). Even so, her caste identity was no longer an overriding feature insofar as being a Vaishnava Baul means acquiring religious expertise. In knowing how to moderate the temperature of seed, she and her husband could insist that they were clean insofar as they knew how to purge their minds and bodies of malignant forces. Their spiritual insights were made manifest by the red clothing they acquired during the rite of initiation, garments that Tara likened to the garments worn by Brahmins during their sacred thread ceremony and which in Tara's view symbolizes procreative female seed. In sum, then, there was not a single motivational and emotional force underlying Tara's reason for becoming Baul. Still, her personal history, her views on gender, caste, health, and longevity indicate a pattern which revolves around notions about suffering as a means of achieving mental, emotional, and physical well-being.

The themes of love and suffering evoked by Tara are not limited to her alone, but familiar topics for low-caste non-renouncers too, both women and men. Thus when Muci Ram Das mixed the curries into an undifferentiated substance, people listening to this narrative would recognize his practice as a hardship, resembling the hardships women undergo when neglected or beaten by their husbands, as Tara's mother was. Such an act of hardship also resonates with the hardships people must endure when submitting to behavioral restrictions through fasting or otherwise. Retaining seed is likewise viewed as a form of hardship, an act conceived of as difficult to implement. Like other forms of worship, it is indicative of love, the outcome of balancing hot with cold, achieved by means of self-restraint and tempering one's inclinations, which constitutes a central value for villagers in general, especially among those belonging to the lower castes.

Because Vaishnava Baul renouncers belong to a devotional (*bhakti*) tradition, and since they hold that feelings of affection arise by means of suffering, the act of forging social ties should not come as a surprise. As I have argued, begging is regarded as extremely difficult, a duty that engenders love. From Tara's point of view, the seed that she embodies is imbued with thoughts and feelings that have been perfected, which flows into others when she begs by means of singing. The personal bonds forged with others, who are distant at the outset but who should grow close through the course of interaction, are modeled on the ties between children, parents, and siblings, and they convey a notion not only of shared substances but also of interdependency and social obligations. For Tara, then, making the step to become a Vaishnava Baul renouncer implies engagement with the social world where the spiritual substance she imparts is equated with the thoughts and feelings emanating from her seed.

Acknowledgments

I wish to thank the editors for the insightful comments I received while working on this text.

Notes

1. Chilluri is a pseudonym as are the personal names of the people I portray.
2. Urban points out that although scholars assume that the Baul tradition arose during the sixteenth century, the name Baul does not appear in written sources until the1880s and 1890s. Urban argues that the category Baul appears to be a scholarly construction employed for political ends (1999:14–15).
3. I found that the name Baul had a slightly more restricted meaning than the term Vaishnava. Villagers (Bauls and non-Bauls) use it to denote mendicant singers of Baul songs who had not necessarily taken *sannyas*. Although Openshaw (2002) and Urban (1999) cogently argue that the name Baul is a fairly recent construction, that it did not appear in print until the 1890s, and that it has probably been created by the upper-middle-class elite to symbolize Bengali identity, I have chosen to retain it seeing that the subjects I describe employ it.
4. See Salomon (1991) for details of his life.
5. The word she used was *rup*, which translates as beauty, form, and color.

Bibliography

Alter, Joseph S. 1992. *The Wrestlers Body: Identity and Ideology in North India*. Berkeley: University of California Press.

———. 1997. Seminal Truth: A Modern Science of Male Celibacy in North India. *Medical Anthropology Quarterly* 11(3):275–298.

Bottéro, Alain. 1991. Consumption by Semen Loss in India and Elsewhere. *Culture, Medicine and Psychiatry* 15:303–320.

Capwell, Charles. 1986. *The Music of the Bauls of West Bengal*. Kent, OH: Kent State University Press.

Cort, John. 1999. The Gift of Food to a Wandering Cow: Lay-Mendicant Interaction Among the Jains. *Journal of Asian and African Studies* 34:89–109.

Daniel, E. Valentine. 1984. *Fluid Signs: Being a Person the Tamil Way*. Berkeley: University of California Press.

Fuller, C.J. 1992. *The Camphor Flame: Popular Hinduism and Society in India*. Princeton, NJ: Princeton University Press.

Gold, Ann Grodzins, and Bhoju Ram Gujar. 2002. *In the Time of Trees and Sorrow: Nature, Power, and Memory in Rajasthan*. Durham: Duke University Press.

Gold, Daniel. 1999. Nath Yogis as Established Alternatives: Householders and Ascetics Today. *Journal of Asian and African Studies* 34(1):68–88.

Hanssen, Kristin. 2002. Ingesting Menstrual Blood: Notions of Health and Bodily Fluids in Bengal. *Ethnology* 41(4):365–379.

———. 2000. Å tigge er en plikt: Vaishnava baulers tro og praksis. In *Nærbilder av India: Samfunn, politikk og utvikling, Kathinka Frøystad*. Eldrid Magerli and Arild Engelsen Ruud, eds., pp. 104–128. Oslo: Cappelen akademisk forlag.

Hart, George. 1973. Woman and the Sacred in Ancient Tamilnad. *Journal of Asian Studies* 32(2):233–250.

Khandelwal, Meena. 2001. Sexual Fluids, Emotions, Morality: Notes on the Gendering of Brahmacharya. In *Celibacy, Culture, and Society: The Anthropology of Sexual Abstinence*. Elisa J. Sobo and Sandra Bell, eds., pp. 157–179. Madison: University of Wisconsin Press.

Laidlaw, James. 2000. A Free Gift Makes No Friends. *Royal Anthropological Institute* 6:617–634.

Lamb, Sarah. 2000. *White Saris and Sweet Mangoes: Aging, Gender, and Body in North India*. Berkeley: University of California Press.

Marriott, McKim. 1976. Hindu Transactions: Diversity Without Dualism. In *Transactions and Meaning: Directions in the Anthropology of Exchange and Symbolic Behaviour*. Bruce Kapferer, ed., pp. 109–142. Philadelphia: Institute for the Study of Human Issues.

Openshaw, Jeanne. 2002. *Seeking Bauls of Bengal*. Cambridge: Cambridge University Press.

Salomon, Carol. 1991. The Cosmogonic Riddles of Lalan Fakir. In *Gender, Genre, and Power in South Asia*. Arjun Appadurai, Frank J. Korom, and Margaret A. Mills, eds., pp. 267–304. Philadelphia: University of Philadelphia Press.

Trawick, Margaret. 1991 [1980]. On the Meaning of Shakti to Women in Tamil Nadu. In *The Powers of Tamil Women*. Susan Wadley, ed., pp. 1–34. Maxwell School Foreign and Comparative Studies, South Asian Series, no. 6, Syracuse University.

———. 1992. *Notes on Love in a Tamil Family*. Berkeley: University of California Press.

Urban, Hugh B. 1999. The Politics of Madness: The Construction and Manipulation of the "Baul" Image in Modern Bengal. *Journal of South Asian Studies* 22(1):13–46.

Chapter 5

Staying in Place: The Social Actions of Radha Giri

Sondra L. Hausner

Yogis in South Asia are known as wanderers who live in no fixed location. Wandering is supposed to loosen ties with the material world, and cultivate detachment and religious knowledge in Hindu renouncers. In both religious text and popular legend, yogis are supposed to wander through the countryside for all but the four rainy months of the year, spending no more than one night in a place and eating what food is given. They should possess only a blanket, a water pot, and a staff to ward off wild animals.[1]

During my fieldwork, however, I found that most renouncers do actually base themselves in particular places.[2] On occasion, they depart to visit a pilgrimage destination or to attend a religious festival, but when they do, they leave a base behind, sometimes with a padlock on the door, signifying eventual but certain return. The words for "home" in Hindi and Nepali are certainly shunned by renouncers, who speak instead of their "seats," (*asan*) or their "places" (*sthan*). But yogis and *yoginis* use their seats as home bases, and they interact with members of local communities as long-term and active residents.

In this chapter, I look at the lives of women yogis, or *yoginis*, in the lower Himalayan region of North India and Nepal. I argue that contrary to popular legend, women renouncers do not wander perpetually between holy sites, but tend to settle in communities, and when they do, they often contribute to their new local communities in ways reminiscent of householder women.[3] Although *yoginis* consciously leave behind normative householder

Reprinted with permission from the *European Bulletin of Himalayan Research* (Vol. 28, Spring 2005).

social structures, they still feed people, protect children, and teach religious values in their new roles as renouncers.

Being a woman renouncer, therefore, both reflects and defies women's roles in householder communities. Women renouncers do not necessarily wander to accrue religious power, but they do leave their natal and marital homes and find new places to settle as full-time religious practitioners. And in their new communities, women renouncers use their sedentary seats as bases from which to care for people. They do not reject the world but immerse themselves in it, using the religious power with which they are bestowed for the benefit of others.

One premise of my chapter is that *yoginis* use renunciation as a way to move out of stifling home communities. Once they have moved into new locations as renouncers, they support and are supported by the lay Hindu communities that are now their own. But in a new place, and with the new status (ambivalent though it may be) of renouncer, the caring, feeding work of women has a very different resonance. In their new locations, women renouncers perform these tasks as *yoginis*, rather than as wives.

In the four sections that follow, I try to show how renunciation is simultaneously about departure and stasis, for different social reasons and with different social results. First, I describe Radha Giri, a fiercely independent and staunchly sedentary *yogini* I met during fieldwork. Radha Giri's story clearly shows how staying in place can be translated into social activism for women renouncers. Second, I review the reasons sadhus, or renouncers, are supposed to wander and, by contrast, I describe the sedentary choices made by the renouncers with whom I worked. Third, I look at the ways renouncers use staying in place to be active in their communities. And finally, I look at how *yoginis* depart from their householder roles—literally and metaphorically—and recreate them in new settings.

Fierce, Sedentary Radha Giri

I met Radha Giri at the 1998 Kumbh Mela, a massive Hindu religious festival, in Haridwar, Uttar Pradesh, and I visited her often when I returned to Haridwar in 2000.[4] A fiercely independent *sadhvi*, or woman sadhu, she was rumored to have magical powers, and she brooked no disrespect toward or disobedience of the rules she had established around her small quarters on the riverbank.[5] Radha Mai, or Mataji, as she was more commonly known, had lived in a tent on a small island in the Ganges River for almost 25 years. Haridwar is one of the most popular pilgrimage places in South Asia,[6] and pilgrim traffic is heavy along the river, which passes directly in front of Radha Mai's place.

Radha Mai was well known for her feisty temper—I heard she was a witch—and also for the unforgiving manner with which she ruled her small stretch of Ganges riverbank. She was a well-respected figure among members of the Haridwar renouncer community (including the men), and among members of the poor local community, who made their living selling trinkets to visiting pilgrims. A steady flow of both Haridwar residents and local and traveling renouncers passed through her place, and I used my visits to her tent as a way to meet other yogis, hear the pilgrimage stories that both sadhus and lay Hindus were eager to tell, and watch pilgrims pay their respects to Radha Giri's place and to the river.

Mataji was neither particularly interested in my interview questions nor particularly verbal, but she was welcoming. I soon learned, however, that when people acted in ways she found inappropriate or disrespectful, she would become enraged and lecture them in a tirade on how to behave properly in the future. On one occasion, I watched her shout at pilgrims who were treating a child unkindly—she dashed out of her tent with her arms flailing, screaming that such behavior was patently unreligious and that the perpetrators should never come near her place again. I learned quickly to follow her instructions, and closely adhere to the code of proper behavior she demanded at her place on the riverbank.

Radha Giri was reticent about her background, a tendency that was quite common in my conversations with renouncers. But over time I did learn that she had been raised and married in the Himalayan area of Kumaon. She had left her marriage—I wondered if her fiery character had contributed to her unwillingness to play the part of subordinate wife—and followed a guru to Haridwar, where she had lived on the riverbank ever since. She was clearly motivated by religious duty, for she unfailingly paid her daily homage to the river and meticulously maintained the altars around the trees under which she lived, although her tent was rather scruffy.

Radha Mai's sedentary quarters were an early indication to me that yogis do not necessarily spend all their time wandering. Mai's style also showed me what staying in place affords renouncers, and how they use the religious power accorded them by local communities for the benefit of those communities. Mataji sat guard over her island, and she used her power as a local *yogini* to make sure that people whom she saw as powerless were properly treated.

Living in one place was the way Radha Giri consolidated her power in and for the local riverbank community. Her social protection efforts were usually directed to women, children, and dogs—creatures in the most need of defense. She mothered the children of the neighborhood, scolding them gently when they misbehaved. She provided work and meals to a madwoman from the area. And she provided shelter for most of the neighborhood dogs, who slept in or near her tent. In one instance, a young man from the area playfully pulled a leaf off the pipal (holy fig) tree under which she lived. Mataji, as protector of place, sternly but patiently explained that he must never disrespect a holy tree, especially in the presence of Mother Ganga. And on more than one occasion, she wrathfully drove away men whom she felt were questioning me too eagerly, telling them that unless they came for reasons of religion, they were not to come at all.

Most notably, since about 1996, Radha Giri has reared a small girl child, whom she named Ganga Giri. The story goes that she found the newborn baby floating down the Ganges in a banana leaf basket, and saved her life. The mythical rendering of how she came to raise the abandoned child deliberately refers to the river that flows in front of her place. The child arrived in her domain, buoyed by the sacred waters of the Ganges, and Radha Mai had no choice but to take her in. She refused to hand the child over to state authorities who wanted to take Ganga to an orphanage—the baby had arrived in front of her doorstep, or tentflap as the case may be, and she would care for it.

Mai was a highly unusual mother figure, to be sure, closer to the age of the child's grandmother than its mother, and more concerned with the bare survival of the child than with any kind of long-term planning for school or for marriage. Feeding the child and tending to her medical needs seemed to be all Mai could afford or cope with. But as a woman, she told me, she felt responsible for the people around her, and all those powerless creatures who found their way to her tent, be they lost anthropologists or helpless infants.

Radha Giri was fiercely sedentary, as she was fierce in most of her actions. Her seat in the small tent was by a sacred fire, or *dhuni*, beneath two large trees and across from the sacred river. Since the time she took over this spot from a different renouncer who had vacated it, thereby becoming the resident *yogini* of the area, she had not lived away from the island. She had refused to leave even during the massive Kumbh Mela festivals in Haridwar, when state authorities asked her to move into the sadhu camps in the city proper, with the other women renouncers. "Once," Mai defiantly told me, "I moved upriver." But that was for two weeks only, and apart from that single occasion, the place between the two trees was her seat, and she would not budge.

Moving through Space and Staying in Place

Renouncers who live in one place disrupt one of the most popular images of asceticism. The Samnyasa Upanisads propose strict guidelines for the real ascetic: no more than one night must be spent in a village, no more than three in a town, and no more than five in a city.[7] These rules seem carefully calibrated to population size: they demand a very low ratio of nights per place in order to prohibit extended social contact, which might lead a renouncer to become attached to the people who live in a particular place. Place is one of the primary ways people become attached to one another, the texts suggest, and wandering is designed precisely to remove the threads of social connection.

The image of the wandering renouncer is powerful because it implies that yogis leave places where householders live in illusion-filled homes. The symbolic act of wandering insists that sadhus have broken free from the spatial constraints of social life. Since they have no place in which they are rooted, and no location through which they are governed or socialized, wanderers live outside the social fray.[8] Wandering has always been part of sadhu life: the excitement and freedom of travel is one of the prerogatives of what for many is an otherwise difficult life choice.

Real-life renouncers are different from textual ideals in many ways, but the ideal of the wandering sadhu has had a particularly firm grasp in the public imagination. Even authors who have been instrumental in pointing out how renunciation does not fit into textual models—like Kirin Narayan, who eloquently deconstructs even that famed opposition between renunciation and caste society (1989)—emphasize the idea that householders are sedentary and renouncers are not.[9] But almost all renouncers I met—both men and women—were sedentary, and many told me that moving around was plainly counterproductive, since it distracted them from regular religious practice.

The circumstances of sedentariness varied among the renouncers I met. Some, like Radha Giri, moved to the places where their gurus had lived. Others found an ashram in a holy place, where they felt protected from the very difficult householder lives they had left behind. Some renouncers lived in a particular temple; some lived in a small room or *kuti*, which was affiliated with a temple or an ashram; many lived at a *dhuni*, or sacred fire pit. Traditionally, I was told, a sadhu should sleep under a tree or in a temple, but the yogi seats that I saw varied widely: caves, ashrams, *dhunis*, hotel rooms, apartments, tea shops, riverbanks, huts, tents, or *kutis*, in residential courtyards or temple complexes, made of stone, concrete, straw, brick, or wood, all served as bases for sadhus I knew. Most renouncer seats waved the small triangular red flag that also flies from temples, a public symbol of religious activity.

A yogi's place is certainly not a *ghar*, however, which is an exclusively householder term for house or home in both Hindi and Nepali. My informants used the word *asan* to refer to the specific places where they lived. From the Sanskrit verb *ās-*, to sit, to stay, or to live, *asan* means seat. Renouncers' seats—sometimes literally marked by a small, portable rug or deerskin—are the places they stay unless they are traveling to festivals or moving on pilgrimage. *Asan* refers both to an external seat and to the internal seat or balance of the body. (Physical yoga postures are called *asanas*, because yogis are instructed to use their bodies to maintain the steadiness of a pose.) Through this language, renouncers differentiate their bases and places from those of householders: their dwellings are not homes, but mobile places of meditation.

Many renouncers choose to live in pilgrimage places, since these are holy locations infused with both a history of powerful religious practice and an infrastructure that will provide material sustenance, given a steady flow of pilgrims ready to support religious practitioners. Any place holy and convenient could be a sadhu seat, and sometimes a good spot recently vacated would be quickly reoccupied by another sadhu. When I asked Radha Giri why she had chosen to live in Haridwar, she thought it was a ridiculous question. "Where do you want me to live?" she retorted. "The railroad station?" The idea that a yogi should be transient rather than live within a community (or that a renouncer would choose a particular place, rather than accept her karmic destiny) seemed absurd in her view. Transience ran counter to her core ideas about the purpose of renunciation.

Meditation through Action

Despite the popular image of the wandering yogi, there is a long-standing tradition of respect for renouncers who stay in one place. Staying put means yogis can do their religious practice (*sadhana*) and publicly receive and bless pilgrims. The places where renowned saints live usually take on religious significance. Two renouncers with whom I lived in the Ganges Valley for a few days told me with pride how their Rishikesh-based guru never left his ashram—or even his cave—while he was alive, even to cross the bridge into the main town. Certainly pilgrims who visited Radha Giri's small island paid homage to the altars under her trees without fail. One local man offered incense to the river and to Mai's altar every evening.

Some of the sedentary women renouncers I met wished for isolation, wanting to do their religious practice undisturbed. But for others, staying in place meant that rather than detach from society, they became active, vigorous participants in it, on their own terms and in their own ways. Popular legends about yogis explicitly refer to their community participation—renouncers are known as healers, religious storytellers, and compassionate ritualists. As Narayan writes, "Ironically, the act of renunciation may in fact push an ascetic into more extensive social involvement than if he or she remained a layperson" (1989:74).

In Radha Giri's interactions with the local community, we see exactly the kind of social engagement that the texts caution against: relationships, emotional involvements, parenthood, and an ongoing system of exchange between a renouncer and the residents of a particular place. A Hindu renouncer's religious practice is supposed to focus on liberating the Self, not improving society. Yogis are supposed to be socially detached, not socially engaged. But Radha Giri interpreted her religious practice as a kind of community activism. Her fiery judgements of both pilgrim and resident behavior were a way of protecting people—particularly the downtrodden—in her area. Shouting at householder men—an action that might be written off as *yogini* madness—was, for Radha Giri, a way to protect a woman or child who was treated badly.

In turn, Mai was accepted by the local community, and was allowed and encouraged to remain sedentary. (By contrast, one yogi I knew was run out of town for improperly treating local women in the community in which he lived.) Radha Giri's *dhuni* was treated as a safe and holy place, where children came to play and neighborhood residents came to offer incense and receive blessings during the evening rituals. By supporting the community through her protective behavior, she was entrusted with religious power, which she in turn used to make sure people were treated well. Her social efforts were place-specific, even though as a renouncer she was precisely supposed to avoid staying in place. She held her place in the community by actively participating in it.

Radha Giri was a rather unique *yogini*, but the way she cared for people in a single location was reflected by many other *yoginis* I met or heard about. A *sadhvi* named Tapovan Ma abandoned her solitary ascetic practices at the 14,000-foot base camp of the mountain Shiv Ling when her health started to fail, and doctors encouraged her to live at a lower and warmer altitude. She moved to a town a few hours' drive down the mountain valley, and changed her severe, isolated *tapas* (austerities or yogic discipline), into a practice of feeding people daily. A Western renouncer I met used her ashram home-base to feed the entire local community of renouncers each evening. Another raised funds to create an ashram high in the mountains where renouncers who needed a place of respite could continue their religious practice and study. (This woman was actually concerned that by creating an ashram and living a sedentary life, she was going back on renouncer ideals.) A close informant in Kathmandu was reared on the grounds of the Pashupatinath temple by her *yogini* grandmother, who had refused to let a baby granddaughter go uncared for when her mother died.

All these renouncers used their power as respected local *sadhvis* to make sure people were treated well. Over the course of my fieldwork, many renouncers described religious practice to me as something that had to be done completely alone. But community involvement seemed to fulfill a different kind of religious mandate. These sorts of actions were referred to "*kriya yoga*," meditation through action, and were clearly seen as an appropriate *sadhana* for sedentary sadhus (see also Khandelwal, this volume). Both pilgrims and yogis explained to me that a genuine concern for the social good reflects a religious character, and that active social participation proves a yogi's capacity in a way different from but equivalent to solitary ritual practice.

New Places, Old Actions

The *yoginis* I worked with were ascetics who had left their home communities. But they continued to act out women's roles by devoting themselves to caring for others. In their new locations, however, and in their new stances as renouncers, womanly actions were transformed. Caring for children and feeding families no longer took place in the context of subordinate and required behavior, but as the voluntary practice of powerful community women. The actions of care became part of a religious practice, rather than an unquestioned social role.

More than 20 years ago, A.K. Ramanujan wrote that a maternal nature may be seen as contributing to saintliness (1982). Recently, Meena Khandelwal has written on how *yoginis* fulfill the roles of ideal mothers, by both gently scolding and heartily feeding their disciples and visitors (1997, 2004). This construction means that women have an added advantage as renouncers, in a sense: "feminine" qualities naturally provide the love, kindness, and caring nature expected of a saint. The holy Ganges River is Ganga Ma; the revered land of India is Bharat Ma: through their nurturing qualities, the earth and the water become mother figures. And in these constructions, being a mother is the most sacred thing you can be.

Male renouncers also feed people on occasion, and in doing so, they too project the motherly qualities of a highly realized, compassionate being (Narayan 1989; Ramanujan 1982). One man with whom I worked took care of the children of the temple area in which he lived, giving them small amounts of money for completing little tasks. Both yogis and *yoginis* earn the respect of local communities when they help out, and mitigate negative images of the mad and isolated renouncer.

Radha Giri became the literal "Mataji" when she began to raise Ganga Giri. And yet as a loud and opinionated woman who smoked hashish, Mai defied the images of both traditional mother and beatific saint. She was unruly and impetuous, and she was widely credited with having real spiritual power, even by those Haridwar residents who told me that most sadhus were nothing more than money-hungry louts. Radha Giri was highly respected because she combined the most powerful qualities of a *yogini*: she was both strong and forceful and, at the same time, maternal and caring. She was at once ready to challenge the stereotypes of womanly behavior and ready to fulfill a woman's roles.

The final point I want to make about women renouncers' community participation is that even though their efforts are specifically local, they do not occur in *yoginis*' "native" places. Women renouncers engage in womanly activities—even having left householder society—precisely because they are not in their home places. A woman's place, for Hindu *yoginis*, is in somebody else's home. Every single woman renouncer I met had left the place where she was born or where she was married.

For *yoginis*, choosing to become a renouncer allows departure. Radha Giri's reticence on the topics of her girlhood and marriage was typical of renouncers; although it was rarely explicit, the women I spoke with strongly implied that they became *yoginis* as a way to leave their natal or marital homes. They became *sadhvis* because they wanted to live religious lives, to be sure, but also because they did not want to marry, or because they were unhappily married, or because they were widows and did not know where to turn.[10] The wandering that renouncers are supposed to do means that women who become renouncers can physically leave unhappy domestic situations. As *yoginis*, they can leave home, and create a new place elsewhere. And they use their experiences as women who have suffered to preach detachment, and to protect householder women who have not chosen or been able to leave unhappy domesticity.

Householders and Renouncers: A Concluding Comment

The distinction between householder life and renouncer life is central to the identity of the *yoginis* with whom I worked. But scholars who have tried to find a precise, pan-South Asian category with which to distinguish householders from renouncers have failed: each possible theoretical distinction breaks down when faced with the range of actual, lived experiences among sadhus and *sadhvis*. Some renouncer sects do pay attention to caste. Some ascetics are married, and some do raise children. Most yogis are sedentary. There is no one bar on social engagement in renouncers' lives.

Here we have the fluid subjectivity of renunciation at work. What I mean by this is that different parts of a *yogini's* identity come to the fore in the context of different oppositions. The way a *yogini* demonstrates her status as a renouncer—someone who has departed from householder life—depends on her circumstances and varies by context. For the renouncers I met, teaching detachment was important when faced with the petty intrigues of local gossip, while serving food was important when faced with a community's hunger. Sometimes being a *yogini* means behaving like a mother, rather than accept a child's suffering, while at other times being a *yogini* means behaving like a tyrant, rather than accept a woman's ill treatment.

Nonetheless, I argue that the opposition between being a householder and being a renouncer remains a critical component of yogis' lives.[11] No matter how that opposition is articulated, being able to renounce the place and the status of wife or widow is a driving force for women who become *yoginis*. Renunciation allows women to leave circumstances in which they are not happy, and to break free from so-called traditional women's roles.[12] *Yoginis* do make use of those roles—by feeding people, caring for children, and teaching religious values—in their interpretation of religious practice. But renunciation allows for the break, even if it encourages women to do womanly things in new places and new contexts.

Women renouncers use tradition—both so-called traditional women's activities and traditional paths of renunciation—in their religious lives. And yet they simultaneously use renunciation as a way to break free from those overly strict and structured categories, in the ways they leave oppressive situations and choose to participate in new communities. It is really a rather radical act to leave the oppression of widowhood and become a "traditional" *yogini*, or to leave the structure of marriage and raise someone else's child 20 years later. This is social engagement of the highest order, conducted through a practice renowned for its isolation. In these instances, women renouncers can stop being widows, or unhappy or unwilling wives, and start fulfilling women's roles in contexts where they are respected and thanked.

What about the social attachments that renouncers are supposed to avoid at all costs—the reasons that yogis are supposed to wander? This was a real problem for the *yoginis* I spoke with, who told me in the same breath that renouncer life was both completely blissful and terribly, painfully hard. Radha Giri sighed deeply when she answered this question. I had asked her about the difference between men and women renouncers and she responded by referring to the problem of attachment. "For men, it's easy not to get attached," she said. "They have girlfriends, they act how they want to, and there's no problem. But we women, we want to care for the people around us. We start to love them, and then we feel responsible." Later I realized she was probably referring to the circumstances of Ganga's birth, and to the ironic fact that she ended up caring for a child that was not hers. But she had obviously made a choice: attachment would be the cost of staying in one place as a *yogini*, and she would try to transform that attachment to offer what she could—in her hot, brash, not-so-saintly way—to the beings that came to her door.

Notes

1. See Doniger and Smith (1991) and Olivelle (1992) for good examples of textual requirements for renouncers. This imagery occurs throughout classical Indian philosophy and literature.
2. My fieldwork with contemporary Hindu yogis and *yoginis* was conducted in Nepal and India between 1997 and 2001. I worked closely with three individual renouncers, two women and one man, at their bases in Kathmandu, Haridwar, and Allahabad. In addition, I spoke with hundreds of renouncers over the course of my travels to multiple Himalayan pilgrimage sites and important festival occasions.
3. Elsewhere, I argue that the example of renouncers shows us concretely how communities have traveled across space for thousands of years: that is, modernity and globalization are not solely responsible for the movements and migrations of populations (Hausner 2002a). This essay should not be taken as an attempt to insist on sedentariness for anyone, but rather to fill in ethnographic gaps about how renouncers in South Asia really live: sometimes on the road, and sometimes in place.
4. The two northernmost regions of Uttar Pradesh, Garhwal and Kumaon, together became the new state of Uttaranchal on November 9, 2000.
5. See Lamb (2000) on the powers of women who fall outside of householder norms. Also see Lochtefeld (1992) for another account of Radha Giri's charisma.
6. See Bhardwaj (1973) for an informative breakdown of pilgrimage statistics. Haridwar is popular in part because it is the entry point for the Char Dham, the four Himalayan sites that are the sources of the Ganges and Yamuna rivers.
7. See Olivelle (1992) for the details of *sannyasi* sleeping requirements, and the way they shift over time in the classical literature.
8. See Freitag (1985) on how wandering renouncers may have frustrated British colonial officers, who could not govern people unrooted in space.
9. Gross is the only ethnographer who emphasizes that many sadhus "are part-time itinerant wanderers having some sort of semi-permanent residence from which they make a number of pilgrimages throughout the year" (1992:126).
10. See Arthvale (1930) for a historical account of Hindu widowhood, and Lamb (2000) and Wadley (1995) for contemporary ethnographies.
11. With this argument, I reinsert Dumont's opposition into the renouncer/householder debates (1980[1966]), countering Heesterman's arguments that

renunciation developed as the logical extreme of Brahmanical Hinduism (1982, 1964). But while I argue that fieldwork supports Dumont's fundamental opposition between householders and renouncers, I agree with the many scholarly criticisms that Dumont's system is overly static (cf. Gellner 2001; Das 1982[1977]) and that his emphasis on the renouncer as "individual" is misplaced (cf. Mines 1994).

12. See my AAA paper (2002b) on how a caste critique for renunciation is too limited. Renunciation clearly contains a feminist critique as well, one that we have tried to elucidate in this volume.

Bibliography

Arthvale, Parvati. 1930. *My Story: The Autobiography of a Hindu Widow*. Justin E. Abbott, trans. New York: Putnam.

Bhardwaj, Surinder M. 1973. *Hindu Places of Pilgrimage in India: A Study in Cultural Geography*. Berkeley: University of California Press.

Burghart, Richard. 1983a. Renunciation in the Religious Traditions of South Asia. *Man* (n.s.) 18:635–653.

——.1983b. Wandering Ascetics of the Ramanandi Sect. *History of Religions* 22:361–380.

Das, Veena. 1974. On the Categorization of Space in Hindu Ritual. In *Text and Context: The Social Anthropology of Tradition*. Ravindra K. Jain, ed. Philadelphia: Institute for the Study of Human Issues.

——.1982[1977]. *Structure and Cognition: Aspects of Hindu Caste and Ritual*. 2nd edition. Delhi: Oxford University Press.

Denton, Lynn T. 1991. Varieties of Hindu Female Asceticism. In *Roles and Rituals for Hindu Women*. J. Leslie, ed., pp. 211–231. Rutherford, NJ: Fairleigh Dickinson University Press.

Doniger, Wendy, and Brian K. Smith. 1991. *The Laws of Manu*. New York: Penguin Books.

Dumont, Louis. 1980[1966]. *Homo Hierarchicus: The Caste System and Its Implications*. Chicago: University of Chicago Press.

Durkheim, Emile. 1995[1912]. *The Elementary Forms of Religious Life*. New York: Free Press.

Freitag, Sandra B. 1985. Collective Crime and Authority in North India. In *Crime and Criminality in British India*. A.A. Yang, ed. Tucson: University of Arizona Press.

Gellner, David N. 2001. *The Anthropology of Buddhism and Hinduism: Weberian Themes*. New Delhi: Oxford University Press.

Ghurye, G.S. 1995[1953]. *Indian Sadhus*. Bombay: Popular Prakashan.

Gold, Ann Grodzins. 1988. *Fruitful Journeys: The Ways of Rajasthani Pilgrims*. Berkeley: University of California Press.

——.1992. *A Carnival of Parting: The Tales of King Gopi Chand and King Bharthari as Sung and Told by Madhu Natisar Nath of Ghatiyali, Rajasthan, India*. Berkeley: University of California Press.

Gross, Robert Lewis. 1992. *The Sadhus of India: A Study of Hindu Asceticsm*. Jaipur and New Delhi: Rawat Publications.

Hausner, Sondra L. 2002a. Wandering in Place: Body, Space, and Time for Hindu Renouncers. PhD dissertation, Department of Anthropology, Cornell University.

——. 2002b. Hindu Renouncers and the Question of Caste. Paper presented at the American Anthropological Association Annual Meeting. November, New Orleans, LA.

Heesterman, J.C. 1964. Brahman, Ritual, and Renouncer. *Wiener Zeitschrift fur die Kunde Sud-und Ostasiens* 8:1–31.

——. 1982. Householder and Wanderer. In *Way of Life: King, Householder and Renouncer (Essays in Honor of Louis Dumont)*. T.N. Madan, ed. New Delhi: Vikas.

Khandelwal, Meena. 1997. Ungendered Atma, Masculine Virility and Feminine Compassion: Ambiguities in Renunciant Discourses on Gender. *Contributions to Indian Sociology* 31(1):79–107.

——. 2001. Marriage, Sexuality, and Female Agency at the Site of Sannyasa. Paper presented at the American Anthopological Association Annual Meeting. November, Washington DC.

——. 2004. *Women in Ochre Robes: Gendering Hindu Renunciation*. Albany: State University of New York Press.

Lamb, Sarah. 2000. *White Saris and Sweet Mangoes: Aging, Gender, and Body in North India*. Berkeley: University of California Press.

Lochtefeld, James G. 1992. Haridwar, Haradwara, Gangadwara: The Construction of Identity and Meaning in a Hindu Pilgrimage Place. PhD dissertation, Department of Religion, Columbia University.

Massey, Doreen. 1994. *Space, Place, and Gender*. Minneapolis: University of Minnesota Press.

Mines, Mattison. 1994. *Public Faces, Private Voices: Community and Individuality in South India*. Berkeley: University of California Press.

Narayan, Kirin. 1989. *Storytellers, Saints, and Scoundrels: Folk Narrative in Hindu Religious Teaching*. Philadelphia: University of Pennsylvania Press.

Ojha, Catherine. 1988. Outside the Norms: Women Ascetics in Hindu Society. *Economic and Political Weekly*, April 30:34–36.

Olivelle, Patrick. 1992. *Samnyasa Upanisads: Hindu Scriptures on Asceticism and Renunciation*. New York: Oxford University Press.

——.1994. Ascetic Withdrawal or Social Engagement. In *Religions of India in Practice*. Donald S. Lopez, Jr., ed. Princeton: Princeton University Press.

Raheja, Gloria Goodwin, and Ann Grodzins Gold. 1995. *Listen to the Heron's Words: Reimagining Gender and Kinship in North India*. Berkeley: University of California Press.

Ramanujan, A.K. 1982. On Woman Saints. In *The Divine Consort: Radha and the Goddesses of India*. John Stratton Hawley and Donna Marie Wulff, eds., pp. 316–324. Berkeley: Graduate Theological Union.

Thapar, Romila. 1979. Renunciation: The Making of a Counter-Culture? In *Ancient Indian Social History: Some Interpretations*. Romila Thapar, ed., pp. 63–104. Delhi: Orient Longmans.

van der Veer, Peter. 1988. *Gods on Earth: The Management of Religious Experience and Identity in a North Indian Pilgrimage Center*. London: Athlone Press.

Wadley, Susan. 1995. No Longer a Wife: Widows in Rural North India. In *From the Margins of Hindu Marriage: Essays on Gender, Religion, and Culture*. Lindsey Harlan and Paul B. Courtright, eds., pp. 92–118. New York: Oxford University Press.

Weber, Max. 1958[1920]. *The Religion of India: The Sociology of Hinduism and Buddhism*. Hans H. Gerth and Don Martindale, eds. and trans. New York: Free Press.

Landscapes of Contemporary Traditions

Chapter 6

Passionate Renouncers: Hindu Nationalist Renouncers and the Politics of Hindutva

Kalyani Devaki Menon

> *To live in the face of death? To live in the world and yet be free from the world? Have we ever fulfilled our duty by sitting in the forest? This is why even renouncers have to help society while staying free of human affairs* [insani batein]. *We have to fulfil our duty/responsibility* [daitva] *not run away from it. We should not run away from problems. Today it is very important for India* [Bharat] *to understand the message of the* Gita.
>
> —Sadhvi Rithambara

Sadhvi Rithambara is a powerful female renouncer belonging to the Sadhvi Shakti Parishad, a branch of the Hindu nationalist movement in India whose membership is limited to Hindu female renouncers (*sadhvis*). She was speaking to a rapt audience of men and women in Ramakrishna Puram in New Delhi in 1999 during the Sri Mad Bhagavad Gita Gyan Yagya. This three-day event was organized by the Vishwa Hindu Parishad (VHP) to spread the message of the Bhagavad Gita, an important Hindu scripture. Rithambara uses the Gita to argue that the principles of worldly involvement and renunciation are not opposed to each other. According to her, renouncers must fulfill their duty to society while also maintaining their detachment from the everyday involvements of human beings. These ideas deviate from the construction of renunciation in classical Hinduism, which suggests that a true renouncer is one who lives in the forest, distant from worldly passions and involvements.

Classical theories of renunciation suggest that the renouncer must remove him/herself from the world of attachments, desires, luxuries, and possessions in order to attain liberation from the cycle of *samsara* (cycle of birth, death, and rebirth). Far from imbricating oneself in the attachments and passions that mark the social and political world that ordinary humans inhabit, the Mundaka Upanishad asserts that *moksha* (liberation) will be attained only by those who abnegate involvement in the world and live a life of "penance and faith" focused on penetrating the illusions of the world (Olivelle 1998:270). Yet, how should we understand the deep involvement of female renouncers in the political battles of Hindu nationalism? Rather than turning to the Upanishads to define their place and role in the world, Hindu nationalist *sadhvis* use the Bhagavad Gita to suggest that it is not only possible, but in fact imperative for them to engage in righteous action that upholds dharma, or the moral order of the world.[1] As I will show, these female renouncers also argue that it is the sacred duty of Hindus to participate in this struggle for dharma identified as the struggle to establish India as a Hindu nation.

In this essay I examine how Hindu nationalist female renouncers use religious texts and ideology to disseminate the politics of the movement as well as to justify their own worldly involvements. Through an examination of speeches and conversations with Hindu nationalist renouncers collected while conducting fieldwork among Hindu nationalist women in New Delhi in 1999, I analyze the way female renouncers interweave religious ideas with the politics and agendas of Hindu nationalism (Hindutva). Their efforts not only imbue political action with a sacred injunction but also suggest that their calling is not to remove themselves from the world but rather to shape that world and infuse it with the values, ideology, and politics of Hindutva. Hindu nationalist renouncers skillfully use religious themes to mobilize people to participate in the politics of Hindu nationalism. I argue that it is their ability to weave the political imperatives of Hindu nationalism with the realm of the sacred that makes renouncers such effective spokespersons for the movement.

While most Hindu nationalist *sadhvis* do not make the headlines in India, some like Sadhvi Rithambara and Uma Bharati are among the most prominent figures in the movement (figure 6.1). It is the strident voice of Sadhvi Rithambara that is instantly familiar to those within and beyond the movement, eclipsing the Hindu nationalist male renouncers. In this essay, I explore the religious, and often highly gendered, imagery used by female renouncers in the movement as they give voice to the political agendas of Hindu nationalism. I suggest that it is their dual status as renouncers and as women that makes these *sadhvis* such effective voices for the values, morals, and politics of the movement.

Figure 6.1 Uma Bharati, 1990
Source: Photo by Meena Khandelwal.

While conducting fieldwork I had the opportunity to attend and record several *pravachans* (religious lectures) given by different female renouncers who belonged to the Hindu nationalist movement as well as to analyze the ethnographic context in which they were uttered and made meaningful. The *pravachans* enshrined the political aspirations of Hindu nationalism thus reflecting the ways in which the sociopolitical world shapes, inflects, and indeed engenders religious ideas (Asad 1983). Each of these *pravachans* addressed the political agendas of the movement in different ways, but the lectures varied in the emphasis given to ongoing political debates over religious themes. Some only tangentially referred to religious themes, while others focused explicitly on religious ideas and used them to interpret the political world and frame activism within it.

At these events, it is the audience's acceptance of these women as religious renouncers, and as people who have the wisdom to penetrate the everyday illusions of the world, that gives their words added power and efficacy. For those who listened to them, *pravachans* carried the weight of a higher spiritual authority precisely because they were delivered by renouncers who are seen to transcend the pettiness of everyday life and to embody the moral vision the movement claims to infuse into Indian politics. As socially engaged renouncers who view service to society as a part of their spiritual practice, Hindu nationalist *sadhvis* situate themselves within a long tradition of modern renunciation that includes famous individuals such as Swami Vivekananda, Swami Chinmayananda, and Mahatma Gandhi. As I argue later, like Hindu nationalist *sadhvis* today, Vivekananda, Chinmayananda, and Gandhi also infused their religious ideas with the nationalist politics of their particular historical moments. Yet the passion with which Sadhvi Rithambara gives voice to the aggressive and violent politics of Hindu nationalism distinguishes her from these renouncers. Below I examine the gendered religious politics of Hindu nationalist *sadhvis* in contemporary India. After a brief discussion of Hindu nationalism, I first look at the ways the Bhagavad Gita offers an ideology of action. Then I show how invoking religious texts is a way to encourage participation—even violence—in the struggle for a Hindu nation. Finally, I look at the degree to which the movement relies on *sadhvis* to mobilize Hindus to engage in the violent politics of Hindu nationalism.

Hindu Nationalism and Religion in Contemporary India

Hindu nationalism is a movement consisting of many different organizations all of which share the common goal of establishing India as a Hindu nation. While the movement dates back to the early part of the twentieth century, particularly with the establishment of the Rashtriya Swayamsevak Sangh in 1926, it has changed from being a voice on the margins of Indian politics to one that has dominated the sociopolitical landscape of the country since the 1980s. From March 1998 until May 2004, the Bharatiya Janata Party (BJP), the electoral wing of the movement, formed the national government in India, albeit in coalition with other political parties. Although based in India, the Hindu nationalist movement also has a significant global presence. The Hindu diaspora is a major supporter and funder of Hindutva and has brought many Hindu nationalist leaders to the West.

There are several different organizations that are part of the Hindu nationalist movement; I mention only a few. The Rashtriya Swayamsevak Sangh (RSS) forms the core of the movement and many of the members and the leaders of the other Hindu nationalist organizations have their roots in the RSS. The RSS is an all-male voluntary paramilitary organization that has an estimated membership of anywhere from 2.5 to 6 million followers (Bhatt 2001:113). The BJP was founded in 1980 and, as mentioned above, has played an important role in Indian politics. Its first president, Atal Behari Vajpayee, served as the prime minister of India from 1998 until 2004. The senior leadership of the BJP continues to maintain close ties with the RSS.[2] The Vishwa Hindu Parishad (VHP) was formed in 1966 and is an organization committed to the revival of Hindu religion and culture. It was at the forefront of the Ramjanmabhumi movement to destroy the Babri Masjid in Ayodhya, a mosque that allegedly stood at the place of a temple marking the birthplace of the Hindu god Ram. It was during this movement that Hindu renouncers became prominent in the cultural politics of the movement using their renunciant authority to rouse Hindus to destroy the mosque and build a temple in its place.

While both male and female renouncers joined the movement, it is important to note that female renouncers, often less visible in traditions of renunciation (Khandelwal 2004), became quite prominent during the Ramjanmabhumi agitation. The VHP established the Sadhvi Shakti Parishad (Organization of Sadhvi Strength) to create a platform through which Hindu female renouncers could play a vital role in the politics of the Ramjanmabhumi movement. In particular, two female renouncers, Sadhvi Rithambara and Uma Bharati, escalated to fame as they stood in front of crowds of Hindus exhorting them to defend Hinduism, destroy the mosque, and kill Muslims, who were constructed as the paramount threat to Hindu India (Sarkar 2001:269). Scholars have noted the highly sexual and gendered imagery employed by these *sadhvis* (Ghosh 2002; Basu 1995) in their condemnations of Muslim men. In their speeches they suggest that Muslim men committed atrocities against Hindu women and figuratively violated the honor of India conceived as the Goddess Bharat Mata. The power of women (albeit renouncers) calling on men to avenge the honor of Hindu womanhood is central to the successful mobilization of men to engage in acts of violence against Muslims. The highly visible and public role *sadhvis* play in Hindu nationalist politics puts in sharp relief, perhaps quite intentionally, the movement's portrayal of Islam as a religion that oppresses women. While most Hindu nationalist ideologues are men, it is important to understand how the agency (Mahmood 2004; Ahearn 2001) of *sadhvis* is critical to constructing the affective power of Hindu nationalism in the minds of ordinary people.[3]

One cannot, however, view the Hindu nationalist movement as a monolith with a singular ideology and agenda for political action. In fact there are opposing tendencies and tensions within the Hindu nationalist movement.[4] Rajeev Bhargava argues that what keeps the various wings of the movement united are, among other things, the desire to establish India as a Hindu nation and a hatred of Muslims, Christians, and secularists (2003:11). In my interactions with Hindu nationalists in New Delhi, I found that while they agreed that India was a Hindu nation and while they perceived and treated Muslims and Christians as "other," not all in the movement hated those belonging to these minority religions. However, whether or not they hated minorities, most of them suspected that the loyalties of Muslims and Christians lay elsewhere. The border dispute between India and Pakistan was a great concern among those I worked with in 1999, especially after the discovery that the Pakistani army had crossed the line of control in Kashmir and invaded Kargil district in India resulting in the Kargil War discussed below. The presence of Pakistanis on Indian soil strengthened the suspicion that many Hindu nationalists already harbored about the loyalties of Indian Muslims.

The "othering" of Muslims by Hindu nationalists is in keeping with other right-wing movements that, as Paola Bacchetta and Margaret Power assert, are those that "differentially draw on, produce, and mobilize naturalized or culturalized self/Other criteria to reify or forge hierarchical differences" (2002:4). Amrita Basu (1999) asserts that Hindu nationalism is more concerned with using religion as the basis for political identity and to construct the Muslim "other" than in using scripture to comprehensively define Hindu culture and society. The *pravachans* of Hindu nationalist renouncers discussed here are absolutely central to the production of the self and the "other" in the Hindu nationalist movement. *Sadhvis* play a critical role both in establishing a Hindu identity for the nation and in constructing Muslims as the enemy of a Hindu India. Using religious texts to interpret contemporary politics, they engage religious motivations to mobilize Hindus to participate in the politics of Hindu nationalism. In what follows I examine the religious sources used by Hindu nationalist renouncers to justify their worldly involvements in the politics of Hindu nationalism.

Renunciation and Action

But those in the wilderness, calm and wise,
Who live a life of penance and faith,
As they beg their food;

> Through the sun's door they go spotless,
> To where that immortal Person is,
> That immutable self.
>
> —Mundaka Upanishad
> 1.2.11 (Olivelle 1998:270)

Living in the wilderness, practicing various austerities, holding no possessions, and devoting oneself entirely to the experience of oneness with ultimate reality—such is the life of a renouncer according to classical Hinduism. Fulfilling one's duty belongs to those caught in the web of *samsara*, that endless cycle of birth and rebirth that the renouncer seeks to escape. As the above excerpt from the Mundaka Upanishad suggests, the goal of renunciation is to attain *moksha* through the merging of the individual *atman* (soul) with ultimate reality, *brahman*, "that immortal person" (Olivelle 1998:270). The Mundaka Upanishad attacks those who believe that they will attain *moksha* through performing the fire sacrifices that were central to the early Vedic period, saying:

> The fools who hail that as the best,
> Return once more to old age and death. (270)

In other words, those who believe that liberation will be achieved through performing Vedic rituals and elaborate fire sacrifices will only be reborn again, bound as they are to that endless cycle of *samsara*. These words suggest that actions bind one to *samsara*, an idea that is important to the Upanishads. To attain liberation one must enter a state of nonaction. A renouncer is required to stop performing rituals and tending the sacred fire. S/he is also required to enter a state of complete detachment from relationships, from events, and from the world in general to focus on understanding the nature of ultimate reality.

Clearly, Sadhvi Rithambara's claim that renouncers must learn to maintain their detachment from the world of humans and yet continue to serve society contrasts with the prescriptions for renunciation that one finds in the Upanishads. However, while many renunciant traditions in contemporary India continue to emphasize detachment from the social world (Vallely, this volume), the worldly involvements and concerns of Hindu nationalist *sadhvis* are not unusual (Knight, Hanssen, and Gutschow, this volume). In the nineteenth century, Swami Vivekananda established the Ramakrishna Order that required its members to perform *seva* (service) (Elder 1990:32). He argued that renouncers should participate in the religious and cultural revival of the nation and engage in efforts to ameliorate the socioeconomic conditions of the country (Sinclair-Brull 1997:32, cited in Khandelwal 2004:115).

Swami Chinmayananda, an important and respected figure in modern Hinduism, was also deeply involved with the Vishwa Hindu Parishad's early activities to revive Hindu culture in independent India (Jaffrelot 1999: 194–196). More recently, Swami Ranganathananda, the late president of the Ramakrishna Math and Mission asserted that the idea that liberation is gained through inaction has deluded both renouncers as well as householders and "ruined the nation" (Ranganathananda 2000:258). He argued that it is only in the modern era, through the ideas of people like Swami Vivekananda, that people are beginning to understand the importance of action. As Meena Khandelwal contends in her recent ethnography on female renouncers in North India, in the modern period renouncers have often included social engagement as part of their spiritual practice (2004:115–116). Indeed, for Baiji, one of the female renouncers Khandelwal discusses (in this volume), *seva* is central to spiritual practice. As Khandelwal argues, in contemporary renunciation, what is crucial is to maintain detachment from the actions one performs.[5]

Hindu nationalist *sadhvis* view their political activity as *seva* and, like other renouncers, claim that there is religious justification for worldly involvement in texts like the Bhagavad Gita.[6] The Bhagavad Gita became popular during the Hindu reform movements of the colonial period (Flood 1996:124; Chatterjee 1986) and has remained an important vehicle for religious politics and religious actions since Gandhi's use of the text to construct his own philosophy of nonviolent action. In contrast to Gandhi's interpretation, Hindu nationalists argue that according to the Gita worldly involvement—even violent action—in defense of the moral order is justified and can lead to liberation. While in the Upanishads the path to liberation is through a transforming wisdom, the Gita suggests that in fact action is more powerful than knowledge in the pursuit of liberation. Krishna tells Arjuna the following:

> A man cannot escape the force
> of action by abstaining from actions;
> he does not attain success
> just by renunciation.
>
> No one exists for even an instant
> without performing action;
> however unwilling, every being is forced
> to act by the qualities of nature. (Miller 1986:41)

Here Krishna argues that a person cannot escape action. Even a renouncer, who has supposedly entered a state of inaction, continues to perform actions because s/he is human. According to Krishna, renunciation of action does not necessarily lead an individual toward *moksha*. Krishna argues that *moksha* is achieved not through renunciation of action but rather through performing "necessary action" with complete detachment (42), an idea which Hindu nationalist *sadhvis* rely on—explicitly or implicitly—to promote their vision of a unified India.

However, Krishna must now explain the contradiction between the idea that actions tie one to the cycle of *samsara* and the idea, introduced in the Gita, that actions can also liberate one from the cycle of *samsara*. While Krishna accepts that most actions bind you to *samsara*, he posits that sacrificial actions performed simply for preserving the universe, rather than for personal gain, can be liberating. Thus he tells Arjuna to "perform action as sacrifice" (42). Krishna argues that sacrificial actions performed with complete detachment from the fruits of action cannot bind one to the cycle of birth and rebirth. Krishna reminds Arjuna that it is because action is necessary for the universe that Krishna himself must return in successive incarnations to perform actions (44). Therefore, according to Krishna, "wise men should act with detachment to preserve the world" (44).

This last statement is central to understanding the worldly involvements of renouncers today, since Krishna is suggesting that it is their duty to perform actions that lead to the preservation of the world. They must do so with complete detachment, however, or they risk performing action that will carry karmic repercussions. Sadhvi Rithambara thus finds religious justification in the Bhagavad Gita for her claim that it is the duty of renouncers to "help society while staying free of human affairs." In the language of Hindu nationalism, it is not appropriate for a renouncer to search for liberation alone in the forest; rather it is necessary for the renouncer to live in the world while not being of it. Yet, how does a renouncer "help society?" Below I analyze how Hindu nationalist female renouncers interpret necessary action for those who listen to their religious discourses.

Awakening *Sadhvi Shakti*

What is the role of female renouncers in the politics of Hindu nationalism? Building on a common Hindu nationalist construction of a Hindu nation under siege, Sadhvi Kamlesh Bharati, the head of the Sadhvi Shakti Parishad, argued in a speech given at the Sri Mad Bhagavad Gita Gyan Yagya that in addition to attacks on "mothers and sisters" and attacks on Hindu values, in recent years, "two *lakh* [hundred thousand] Hindu girls are converted by Christians and Muslims every year. Our temples are being converted into mosques. If our *sadhvi shakti* [strength] is awakened then these attacks on our dharma [religion] will stop." Her speech was brief and she limited her statements to explaining to those assembled the importance of organizing *sadhvis* as well as other people of faith in the struggle to protect India envisioned as the goddess Bharat Mata [Mother India]. According to her, the role of the Sadhvi Shakti Parishad was to organize programs such as the Sri Mad Bhagavad Gita Gyan Yagya in order to promote Hindu values, culture, and rituals, as well as to teach people to "consider Bharat Mata as a mother and seat her on the throne of the universe."

The vision of India as Bharat Mata has deep implications for the politics of Hindu nationalism. A relatively new form of the goddess, Bharat Mata was first conceived in Bankimchandra Chattopadhyaya's novel *Anandamath* in 1882 (McKean 1996:44). In *Vande Mataram*, a hymn that Bankimchandra composes to the goddess in *Anandamath*, Bharat Mata is variously portrayed as Annapurna who symbolizes the rich past of the nation, as Kali who symbolizes the angered and violated nation under colonial rule, and as Durga who symbolizes the victorious and powerful future of the nation freed from colonial rule (Sarkar 2001:274). Here the goddess is depicted as both vulnerable as well as powerful, violated by the enemy and then empowered by the faith of her devotees. For the protagonists of *Anandamath*, as for Hindu nationalists today for whom *Vande Mataram* remains a daily prayer, both spiritual and political liberation come from devotion to the goddess (McKean 1996:145).

Tanika Sarkar (2001:278) has argued that Hindu nationalists envision the struggle to defend the integrity of India's borders as a religious necessity because an attack on territory constitutes a violation of the sacred body of the goddess.[7] Central here is the image of women's honor that must be protected from violation, an image that is central to Hindu nationalism (Menon 2005) and to nationalist movements all over the world (Menon and Bhasin 1998; Yuval-Davis 1997). Sadhvi Kamlesh Bharati argues that the dishonor of Hindu women and girls must be fought against alongside the greater dishonor of Bharat Mata and the Hindu values that both represent. In the gendered imagery of nationalist movements it is women who bear the burden of national honor and symbolize the cultural and religious identity of the nation.

Sadhvi Kamlesh Bharati asserts that *"sadhvi shakti"* can stop attacks on dharma. *Shakti* can be translated as strength, but it also specifically refers to the female creative principle that women are thought to embody. Although *sadhvis* are supposed to be neither male nor female, they continue to be viewed as women (Khandelwal 2004) and often continue to play roles that are consistent with those of female householders (Hausner, this volume). Given that women bear the unequal burden in Hindu nationalism of both embodying and transmitting Hindu cultural traditions, values, and morals, it is perhaps the femaleness of *sadhvis* that uniquely positions them to prevent attacks on dharma and also to organize events to promote Hindu ideas. Tanika Sarkar argues that Hindu nationalists believe that women are the "custodians" of Hinduism and "that therefore they can respond to a call that comes from the heart of age-old Hindu beliefs" (1995:209). Thus, it is not surprising that it is *sadhvis*, rather than male renouncers, who play such a prominent role in the cultural politics of Hindu nationalism.

The image of India as Bharat Mata also suggests that it is no longer simply the patriotic duty but also the religious duty of all Hindus to participate in the struggle to defend this goddess/nation. This image is powerfully invoked by Hindu nationalist *sadhvis* in their efforts to mobilize Hindus in the political agendas of Hindu nationalism. At a speech delivered to members of the Hindu nationalist movement gathered to honor soldiers who had died in the Kargil War between India and Pakistan in 1999, Sadhvi Rithambara invoked the image of soldiers sacrificing their lives for this goddess/nation:

> Our greatest respects at the feet of those who gave their lives at the feet of mother India. We salute that mother's womb, that mother's lap, that mother's love, her affection, in whose shadow our country's brave soldiers learnt to sacrifice for their motherland. We salute those widows who laid their *sindhuri* [married] nights, their happy days, at the feet of Bharat Mata.[8]

In this passage, both men and women must sacrifice for Bharat Mata, yet these sacrifices are clearly defined according to Hindu nationalist constructions of gender. Here violence is glorified and men are asked to embrace death in defense of Bharat Mata. Women are lauded for their ability to sacrifice the pleasures of motherhood and sex (codified in the reference to *sindhuri* nights) and for teaching their sons to be brave and patriotic. Indeed, as I demonstrate below, references to the sacrifices of both soldiers and their widows greatly affected the audience. The images of a mother sacrificing her son or of a wife sacrificing her husband resonate with a broader cultural glorification of female sacrifice and are powerful ways to recruit women into the movement. Indeed, Purnima Mankekar has demonstrated how similar images of female sacrifice in popular culture appeal to female viewers in India (1999:259–288).

Sadhvi Rithambara, unlike most female renouncers, often incorporates discussions of sexuality in her speeches quite explicitly. Bishnupriya Ghosh describes Sadhvi Rithambara's speeches as "highly sexual and transgendered" moments in which she "adopts the privileges of male speech (sexual aggression, intimate vulgarities, suggestive insinuation)" (2002:270–271).[9] While her status as a renouncer allows her to distance herself from the restrictions on women's performance, it is her femaleness, as I argue below, that empowers her words.

Sadhvi Rithambara is one of the most skillful orators in the movement. As Tanika Sarkar has argued, her "voice seems always almost about to crack under the sheer weight of passion. The overwhelming and constant impression is one of immediacy, urgency, passion and spontaneity" (1993:30). Ghosh argues that Sadhvi Rithambara is one of the key women in the movement who engage in public performances as "instigators of affect and emotion, and as progenitors of non-rational collective identifications" (2002:259). Women's close association with "emotion rather than reason" (Basu 1995:164) in Hindu nationalist imagery is central to why *sadhvis* play such a prominent role in the movement. Although female renouncers are supposed to have transcended gender (Khandelwal 2004), it is because they are identified as women that they are able to forge the patriotic ardor that drives Hindu nationalism. As a renouncer, Sadhvi Rithambara is supposed to be detached from passions. However, as a woman she is expected to be emotional. Arguably, Sadhvi Rithambara's prominence in the movement is related to the ambiguity of her own subject position as a passionate renouncer.[10]

This production of affect and emotion was central to Rithambara's performance at the Kargil event. The large auditorium of the Constitutional Club on Rafi Marg in New Delhi was filled to capacity, and yet the room was quiet as Sadhvi Rithambara's voice trembled with emotion, and tears streamed down her face. Several in the room wept along with her, caught in the spell of her powerful oratory. As she continued her speech her voice became more angry and passionate until she was almost screaming the words to a silent auditorium: "on Islamabad's chest will rest India's flag, up till Rawalpindi and Karachi . . . up to the Indus river it will all become India.[11] Then for eons and eons there will not be a devil like Jinnah.[12] There will be a Kashmir but there will not be a Pakistan."[13] As the audience cheered in response to these statements, it became clear that Sadhvi Rithambara's passionate anger inflected the imaginations of those present and perhaps recruited them into the discursive constructions of Hindu nationalism. Purnima Mankekar asserts that it is important to recognize the processes through which individuals come to view themselves as subjects within specific ideological formations (1999:17). At this event, the affective power of Rithambara's oratory clearly resonated with those present, creating an arena within which individuals could embrace the politics of the movement and envision themselves as subjects within its particular ideological formations.

The image of Bharat Mata torn asunder by Partition whose sacred body will be reintegrated in a united, and Hindu, India is important in

Rithambara's speech. The idea of *akhand* Bharat (united India) is central to Hindu nationalists and is an image that is called upon repeatedly in speeches given by *sadhvis*, including Rithambara. She refers to this at the Sri Mad Bhagavad Gita Gyan Yagya as well, suggesting that it is the sacred duty of Hindus to fight for an *akhand* Bharat. At this event she skillfully integrates Hindu nationalist politics into her exegesis of religious texts to suggest that the struggle for an *akhand* Bharat is a struggle for the very moral order of the world—for dharma. She begins by saying, "the martyrs of Kargil, according to the guidance of the Gita, sacrificed themselves for their country. They gave their bodies for their country. What will we give? We will listen to the *pravachan* and then go home and live surrounded by comforts." In Sadhvi Rithambara's speech, the battle against Pakistan to defend the territories of India was a sacred war, similar to the war of Kurukshetra that forms the backdrop to Krishna's conversation with Arjuna in the Gita. Krishna suggests that it is Arjuna's sacred duty to fight, and in his cosmic form, Vishvarupa, he shows Arjuna that the relatives and teachers that he is so reluctant to fight are already dead. He tells Arjuna that they will die regardless of Arjuna's actions (Miller 1986:103) because their death is necessary to uphold dharma. Krishna tells Arjuna to act as Krishna's "instrument" so that he may "win glory" (104). Referencing this, Sadhvi Rithambara says:

> Bhagvan Krishna says, Arjun, the Kauravas standing in front of you are evil. They endanger dharma. If you do not kill them Arjun then I will. If the enemy does not understand friendship then it is necessary to give them a strong reply. This is the goal of the Sadhvi Shakti Parishad. I want to make a request to Atalji that you had gone there taking a bus with the message of friendship. But that rogue called Sharif did not understand your friendship. Now it is necessary, instead of taking a bus to take a tank and go. In Pakistan go up to Rawalpindi, Lahore, and Karachi . . . Because we know that those who don't make mistakes we call god. Those who make mistakes and repent we call humans. And those who make mistakes and don't repent we call the devil. But those who make mistakes again and again without repenting we call Pakistan.

The audience began to cheer and clap loudly, clearly delighted by Sadhvi Rithambara's indictment of Pakistan. By juxtaposing the Gita and the Kargil War, Rithambara suggests that both are dharmic wars. In the Mahabharata the Pandavas attempted to prevent war with the Kauravas by trying to convince Duryodhana to concede their share in the kingdom.[14] Similarly, according to Rithambara, Atal Behari Vajpayee, then the Indian prime minister, in an act of reconciliation had taken a bus trip to Pakistan to inaugurate the newly established bus route between New Delhi and Lahore. A few months later India and Pakistan engaged in the Kargil War after the Pakistani army invaded Kargil district in India. Rithambara sees this as reason not only to defend territory but also to fight to regain the dharmic order that, from her perspective, requires an *akhand* Bharat.

I once asked a male member of the movement what *akhand* Bharat included. He informed me that India, Pakistan, Bangladesh, and Afghanistan were all part of *akhand* Bharat. When I expressed surprise at the inclusion of Afghanistan, he responded by saying: Yes, Afghanistan—for after all Gandhari from the Mahabharata is from Gandhar or Kandahar, which is in Afghanistan.[15] Many in the movement believe that all these countries are part of the sacred soil of the Hindu nation, resonating with a common claim of nationalisms that culture, blood, and nation are not imagined but instead firmly rooted in territory. This territory is included in the landscape of the nation through imagined networks forged through folklore—in this case the ancient Hindu epic, the Mahabharata. Also linked to this notion of *akhand* Bharat is the related sense of injustice that this land, firmly linked to the imagination of nation, has been taken away from its rightful claimants. Importantly Sadhvi Rithambara asserts that it is only with creation of *akhand* Bharat that the "river of peace will flow" and that the "sacrifice" of the soldiers will be honored. Until then the intrusions will continue, the violence will continue, and such violence, on the part of Hindus fighting for a moral order, will continue to be justified.

During these events, the affective oratory of Hindu nationalist *sadhvis* becomes central to the movement's ability to frame Hindu nationalist politics within a religious lexicon. Yet, if these framing strategies (Snow et al. 1986) are to be effective, it is necessary to ensure that this religious lexicon can be interpreted in ways conducive to the politics of Hindu nationalism. Below I examine how Hindu nationalist *sadhvis* interpret the message of the Gita for their audiences so that it can become a tool for mobilizing people toward Hindu nationalist politics.

"The Time for the Sudarshan Chakra": Hindu Nationalism and Sacred Duty

In this world if you have to leave acquaintances for the sake of god then leave them. However, don't ever leave god for the sake of acquaintances. If in this world you have to forget acquaintances for god's sake then do so. However, don't ever forget god for the sake of acquaintances. And when a person's mind

> *becomes joined with god then there is no duty that is too much for a person to perform*
>
> —Sadhvi Shiva Saraswati

These words were delivered in a speech by Sadhvi Shiva Saraswati at the Sri Mad Bhagavad Gita Gyan Yagya. This speech, which was heavily laden with Sanskrit verses recited from the Bhagavad Gita and delivered in a monotonous voice, did not raise as much audience participation as Sadhvi Rithambara's. Perhaps the monotonous tone of the speech had something to do with her own irritation at the fact that her entrance had been mistaken for Sadhvi Rithambara's. A man who had been leading the audience in devotional songs before any of the *sadhvis* arrived spotted the entrance of a *sadhvi* and declared energetically into the microphone, "Sadhvi Rithambara *ki jai*" (glory to Sadhvi Rithambara). The audience immediately stood up and began showering her with flowers and saying, "Sadhvi Rithambara *ki jai*." It was a while before either the man on stage or the audience were able to discern Sadhvi Shiva Saraswati's short form through the crowd to realize their error. Ignoring the embarrassed and profuse apologies (also made into the microphone), Shiva Saraswati sat down on the mattress that had been prepared for her at the center of the stage in what seemed like a huff. Neither did she acknowledge the apologetic gentleman nor smile in his direction and instead sat sullenly staring at the audience in front of her. Another young *sadhvi* who had come in with her began to sing a devotional song to the sound of an electronic keyboard that she was playing. During this song, I began to talk with one of Sadhvi Kamlesh Bharati's two young nieces who had escorted Sadhvi Shiva Saraswati in and were now sitting with me in the audience. I found out from one of them that Sadhvi Shiva Saraswati was about thirty years old. Whether it was due to her youthful lack of control over emotion or not, the speech itself, which ran for well over an hour, was monotonous. Yet, although they were not clapping and participating in her speech the way they had during Rithambara's, the members of the audience sat with their hands folded in their laps and seemed to be listening reverently to what Shiva Saraswati had to say.

All the speeches delivered at the event expressed Hindu nationalist politics within a religious framework. Sadhvi Shiva Saraswati's speech focused specifically on sacred duty and attempted to define how Hindus should act according to their duty. She used Krishna's discourse in the Gita to argue that all actions in which a person engages must be undertaken as acts of devotion. She said, "Whatever you do, do it as if it were an act of devotion to god. Now even if you are sweeping the floor or cooking a meal, as long as you are doing this without selfishness and you are doing it as an act of devotion to god, you will show your devotion. If you think of the whole world as god's feet then any act you do in this world is an act of worship at god's feet." Here, her use of domestic roles usually performed by women in a discussion of duty and devotion is important since it reinforces the Hindu nationalist presumption that women's activism within the movement should not challenge their roles within the patriarchal family (Basu 1995).

Having spelled out the need to show one's devotion to god through performing one's duty, Sadhvi Shiva Saraswati asserted that everyone must give a percentage of their earnings to god by "giving to national organizations." She did not name any of the organizations in the Hindu nationalist movement and yet, given that the event had been organized by the Vishwa Hindu Parishad, her statement could be interpreted to mean contributing money to organizations like the VHP. In an interesting reinterpretation of the Hindu *ashramas* (stages of life),[16] Sadhvi Shiva Saraswati said:

> Give a few years of your life to society. When the children get older and start taking care of the business, when you get old, then the mother thinks, the time for *vanaprastha* has come; the father thinks, the time for *vanaprastha* has come. Don't leave everything behind and go off to the jungle.[17] Stay at home. Put all your worry and all your efforts into serving the society that has given you strength.

She repeats these lines again later in her speech after saying that Ram *rajya* (kingdom), Ram's legendary kingdom that Hindu nationalists identify with the establishment of a Hindu India, does not come from speeches but through actions. Shiva Saraswati is not simply telling people to work toward the betterment of society, but rather she is identifying the goal of one's social work to be the establishment of Ram *rajya*, a Hindu nation in India. She asserts that the best way to attain Ram *rajya* is not to "stay at home and admit defeat," but rather to join a "social organization or a national organization." But, she warns, one must perform social work without any ego because that is the message of the Gita. She says, "The life that has ego, it attaches a human. The life that has desire, it attaches a human. The life that brings another misery, that attaches a person. The person who has pride, it attaches a human. But the person who wants to be like Lord Krishna, s/he [*voh*] will be freed from attachments."

Krishna says in the Gita that he engages in action in the world to prevent disorder and the destruction of living beings (Miller 1986:44). He also tells Arjuna that those who perform actions in devotion to him will attain liberation and become a part of him (87). Shiva Saraswati builds on these ideas claiming that if one performs actions like Krishna, for the preservation of dharma rather than for personal gain, one will be freed from all attachments.

She emphasizes that all actions must be performed as acts of devotion to god, and since this entire world is god, anything one does, including participating in the Hindu nationalist struggle for Ram *rajya*, is an act of devotion. Shiva Saraswati suggests that god is more important than any human relationship, and ultimately human relationships should not guide our actions. She makes a role model of those who participated in the struggle for India's independence saying: "Those people who sacrificed their lives to make this country free. . . . On the outside they were fighting a war, they were struggling. But on the inside their minds were focussed on Bharat Mata." According to Shiva Saraswati, wherever duty takes a person, even if it means fighting a war, one must perform it thinking not of oneself or those one is about to destroy, but rather thinking only of how one can serve Bharat Mata.

While Sadhvi Shiva Saraswati expresses the political aspirations of Hindu nationalism, she does not explicitly discuss the Kargil War or prescribe a role for Hindus in the battle. Sadhvi Rithambara on the other hand explicitly prescribes a role for Hindus in the conflict with Pakistan. She ends her speech at the Sri Mad Bhagavad Gita Gyan Yagya saying, "My request to you is when the time is right to play the *bansuri* [flute] then do so. Now it is the time of conflict. It is the time for the *sudarshan chakra* [Krishna's sacred weapon].[18] Join the war. Learn about the circumstances all around you and fulfill your duty." Krishna of Jayadeva's *Gitagovinda* (Miller 1977) is the cowherd who sits in the secret grove playing his *bansuri* and awaiting his lover Radha. This powerful metaphor for devotion to god, where the worshipper is cast as the lover longing for union with his/her beloved (god), is all right, says Sadhvi Rithambara, at certain moments. However, she argues, a time when India and Pakistan are at war is not the moment to sing devotional songs. Rithambara tells the audience that this is the time to pick up weapons[19] and participate in the battle. She continues a few sentences later, "Lord Krishna says: Arjun, those who have to die, they will surely die. . . . Death is simply a change of clothes, a few moments of rest. To live is the greatest art. But this is also the truth. Those who are scared of death have no right to live."

Rithambara ended her speech with these words as the audience burst into enthusiastic applause. She suggests that it is the duty of all Hindus to pick up their weapons and fight in the battle without fearing death. These words, that combine vigilante heroism with religious imagery, clearly appeal to the audience in ways that Shiva Saraswati's speech, drawing as it did more directly from the actual Sanskrit text of the Gita, did not. Yet in both the message is clear: the sacred duty of Hindus is not to live in the jungle and contemplate detachment, it is not to sing devotional songs at gatherings such as these, and it is not, in Sadhvi Rithambara's words, simply "to listen to the *pravachan* and then go home and live surrounded by comforts." In fact it is the religious duty of all Hindus to act, and act violently in defence of the sacred land of Bharat Mata.

What's in a Speech? The Religious Sermons of Hindu Nationalist Renouncers

Hindu nationalist renouncers play a key role in the process through which the political agendas of Hindu nationalism are inscribed with a sacred injunction. If we can agree with Tanika Sarkar's suggestion that communalism is a political movement through which "modern political concepts draw many of their valences from the realm of sacred meanings" (2001: 272), who better to infuse ideas from religious texts into Hindu nationalist politics than renouncers whose task it is to make religious ideas speak to the intricacies of life in the contemporary world? In the landscapes of Hindu nationalism, it is renouncers who become charismatic leaders bridging the gap between the sacred and the mundane. As I have demonstrated above, Hindu nationalist *sadhvis* infuse their religious lectures with the political aspirations of Hindu nationalism. Yet the question remains, how are these powerful acts of oratory viewed by the audience? Below I examine the way in which members of the movement variously responded to the speeches given by Hindu nationalist *sadhvis*, and show how *sadhvis* themselves construe their involvement as exclusively dharmic rather than political.

The speech given by Sadhvi Rithambara at the event organized to honor the "martyrs" of the Kargil War was an eloquent lesson to me about the impact of the words of religious renouncers. The Kargil War took place during the year that I conducted fieldwork in New Delhi, and dominated many of the conversations that I had with members of the movement. I had many conversations with members of the movement angered by the "*ghuspetis*" (infiltrators') presence on Indian soil. There had been reports in the media that suggested that the Hindu nationalist government had ignored intelligence warnings about Pakistani infiltration in Kargil district (Gokhale and Pillai 1999) for months before they actually spoke of it in public and had timed their announcement to appear immediately before the hotly contested national elections. However, the members of the movement I spoke to were convinced that the government was not using Kargil for electoral gains. Wings of the Hindu nationalist movement focused on Kargil in order to mobilize people to support the BJP, the Hindu nationalist political party, during the elections, suggesting that the BJP alone was patriotic enough to defend India's territorial integrity. As the BJP-led government screened jingoistic advertisements on television channels featuring soldiers who had sacrificed their lives for the country, other wings of the movement organized various neighborhood functions honoring those who had died in the battle and honoring the families who had lost one of their own. The Kargil function organized by the Vishwa Hindu Parishad, at which Sadhvi Rithambara gave her speech, was one such event.

While most members of the audience were either members of the movement or people involved in Hindutva grassroots campaigns based at their schools or vocational training classes, a few individuals I met were attending because of personal relationships with members of the movement. Hindu nationalist women play a central role in recruiting other women into the movement. Amrita Basu discusses how Hindu nationalist women, because they straddle the public-private divide, are well suited to bring other women into the fold (1999:179–180). One young woman, Priya Trivedi, who was sitting next to me at the event, was not a member of any of the wings of the movement but had begun attending events at the invitation of Jamuna Sinha, a leader of the Delhi Vishwa Hindu Parishad's women's wing.[20] She told me with awe in her voice that she heard Atal Behari Vajpayee, then the Indian Prime Minister, speak at one such event the previous week. Clearly, her association with Jamuna was giving her access to powerful people, which she may not have had before. In this example, Jamuna's befriending of Priya and her gradual introduction to the movement, illustrates a strategy commonly used by members of the Hindu nationalist movement as well as other right-wing movements in various parts of the world. Kathleen Blee illustrates this process in her analysis of racist women in the United States, where she argues that these women understand the importance of personal contacts when recruiting new members into the movement (2002:133).

Although Priya was relatively new to Hindu nationalist politics she was deeply moved by Sadhvi Rithambara's speech. She had been extremely friendly and vivacious as we conversed before the event began. Her mood was transformed, however, during Sadhvi Rithambara's speech. As Rithambara spoke of the women who had lost their husbands and sons in Kargil, Priya sobbed in the seat next to me. As widows and mothers walked across the stage to receive awards recognizing their loss, Priya continued to cry, empathizing with the loss these women had experienced. Rithambara's words resonated deeply and evoked in her the emotional power of national belonging and national loss.

While the references to religion in Rithambara's Kargil speech were fewer than at the Bhagavad Gita Gyan Yagya, for those in the movement the speech was seen to convey a religious message because it was delivered by a renouncer. At both the events, members of the audience crowded around the *sadhvis* after the speeches to touch their feet and be blessed by them. I had taped all the speeches on my tape recorder and after each of these events, several Hindu nationalist women I worked with asked me for copies so that they could listen to, in their words, the "*pravachan*" again. The choice of language is significant here. A *pravachan* is a religious sermon usually given by a Hindu renouncer. By calling these speeches *pravachans*, the members of the movement I worked with were clearly coding the event as a religious occasion. At the event organized to teach people about the Gita, we were all given "*prasad*" after the prayer led by the *sadhvis* was performed at the end of each day. *Prasad* is blessed food that is traditionally given to devotees at a temple, a religious ritual, or other religious functions. To give *prasad* to those gathered in the audience clearly marks the event as a religious occasion. At these events the political ideals of Hindu nationalism are woven into the religious imaginary of those present.

Another example illustrates the attitude that those attending these events had toward Hindu nationalist female renouncers. At the Bhagavad Gita Gyan Yagya, the audience, including myself, were seated on the ground in front of the stage where the *sadhvis* were sitting. Having never been able to sit cross-legged for any length of time, my knee soon began to cramp and I tried to stretch my leg to relieve the pain. I had barely stretched it out when an older woman immediately smacked my leg and told me crossly that one does not point one's feet at *sadhvis*. Suitably reprimanded, I sat upright, swallowed my pain, and digested the reverence with which the woman viewed the *sadhvis*.

At these events, it is the acceptance of *sadhvis* as religious renouncers, who have distanced themselves from the cycle of *samsara* and who have the wisdom to penetrate the everyday illusions of the world, that gives their words power and efficacy. Amrita Basu asserts that the celibacy of these *sadhvis* is associated with "spirituality, purity, and other worldliness" (1995:161), which makes them powerful voices for the movement's visions. Arvind Rajagopal contends that Hindu nationalist renouncers like Sadhvi Rithambara are powerful precisely because as *sadhvis* their motives are above reproach (2001:232). In other words it is the authority that their religious status as renouncers confers on them that allows them to translate their political zeal into a "pure religious passion" that is supposedly removed from what many in the movement view as the corruption and immorality of politics (Sarkar 2001:286).

A conversation I had with a senior member of the Sadhvi Shakti Parishad about her life and her decision to join the Hindu nationalist movement reveals that Hindu nationalist *sadhvis* construct their own action as dharmic and as distinct from politics. Sadhvi Kiran Bharati became a renouncer at the age of 30. However she told me that since early childhood she had known that she was somehow cut off from the world and interested in spiritual work. In fact, she claimed, nobody who knew her was surprised when she became a *sadhvi*. She joined the movement some years after her decision to become a *sadhvi* at the time of the Ayodhya agitation over the 400-year-old Babri Masjid, which Hindu nationalists claimed was at the site of a temple dedicated to the birthplace of the god-king Ram. Sadhvi Kiran Bharati set off for Ayodhya to find out why Hindus were demanding a temple in the place of the mosque and why Muslims were refusing to allow this. She received the confirmation she needed in the form of the idols of Ram present in the temple, which she felt established the existence of a temple at the site of the mosque.[21] She told me that this made her very angry and she joined the Ayodhya agitation.

The *sadhvi* told me that the VHP had not been responsible for destroying the mosque, since most of its workers were still in Delhi at the time. This statement is contradicted by other VHP women I spoke to in New Delhi. Sadhvi Kiran Bharati explained that the workers of the various wings of the movement had managed to attract a large crowd to the site of the mosque. They had all been sitting around when some monkeys began to play on the roof of the mosque. Given the significance of monkeys in the Hindu epic Ramayana, people were moved by this sight and several rose spontaneously and began to destroy the mosque.[22] She insisted that L.K. Advani and Ashok Singhal, who were prominent in the movement to destroy the mosque, tried very hard to stop them from destroying it but there was nothing to be done about it. However, she insisted, now that the mosque had been destroyed the temple was definitely going to be built.

Sadhvi Kiran Bharati sees her role, and the role of all the *sadhvis* in the movement, as defending Hinduism from perceived attacks from other minority religious communities. Clearly, in her narrative it is not the compulsions of electoral politics but rather the anger or the "pure religious passion" (Sarkar 2001:286) provoked by the sense of injustice to Hindus that motivated her to become a part of the Hindu nationalist movement. She informed me that the Sadhvi Shakti Parishad itself was founded because of this feeling of injustice to Hindus, a sentiment echoed by Sadhvi Kamlesh Bharati's claim that the organization was established to stop attacks against dharma. These narratives are not innocent of political agendas, however, even if *sadhvis* themselves and their supporters place their actions in dharmic rather than political categories. Laura Ahearn argues that language must be viewed as a "form of social action" (2001:110) that actually helps to create the very reality it is supposed to reflect (111). Similarly, in her analysis of women's life narratives, Sarah Lamb contends that narrative, even if inaccurate, must be viewed as "a mode of social action, a creative act of self making and culture making, through the telling of words" (Lamb 2000:20). While we cannot verify whether Kiran Bharati was in fact committed to the politics of Hindu nationalism prior to her visit to the Babri Masjid, her narrative on this point is instructive in the portrait it presents of a renouncer moved by the passion of perceived religious persecution rather than by the seamier world of electoral politics.

By emphasizing her sense of religious wrongdoing Sadhvi Kiran Bharati reveals her own agendas. Among these are first, the desire to present herself as a religious figure and to suggest that it is a threat to dharma that has motivated her to get involved. Second, her narrative legitimizes the destruction of the 400-year-old mosque by claiming, as does the movement, that it was built on the site of a temple commemorating the birth place of Ram after this temple was unjustly destroyed by the Muslim emperor, Babur. Third, she suggests that the destruction of the mosque had divine sanction as evidenced by the presence of the monkeys, symbolizing the monkey-god Hanuman, one of Ram's greatest devotees. And fourth, her representation claims, contrary to evidence and in the light of ongoing legal processes, that the movement, particularly leading figures like Advani and Singhal, was not responsible for the destruction of the mosque. All these are political agendas that are linked to the larger goals of the Hindu nationalist movement. Sadhvi Kiran Bharati's articulation of Hindu nationalist politics as a defense of dharma illustrates the ways in which religious symbols are given meaning through their complex interactions with the workings of power in the world (Asad 1983).

In each of these examples the ability to infuse politics with a sacred sanction, as well as the ability to portray their own political participation as primarily a religious act in defense of dharma rather than one undertaken for electoral politics, arises from the status of these women as religious renouncers speaking from a position of religious authority. To maintain this status it is critical, as Arvind Rajagopal suggests, to promote the idea that these renouncers exist outside the world of politics although they participate in the process of cleansing the political world of corruption and immorality (2001:232). Speaking specifically about Sadhvi Rithambara, Rajagopal argues that she invests renouncers with the responsibility of "cleaning up politics" (232), thus transforming religion "from being the object of action to serving as the means by which the objective is accomplished, and the guarantee that it will be honestly done, given the purity of the leaders and the power of the Hindu tradition" (233).

These ideas need to be situated within the larger context of Hindu nationalist discourse on corruption and politics. Many in the movement believe that the world of politics, narrowly defined as electoral politics and government, is corrupt. The RSS, using this narrow definition of politics, claims that it is not a political organization (Bacchetta 1996:128). This is also a definition subscribed to by many members of the movement—including renouncers—in their attempts to distance themselves from the seamier world of politics (Sarkar 1995:209). Many men in the movement told me that women should not join politics because it was too corrupt. However, women in the movement often disagreed with this assertion. One woman got into a heated argument with a male colleague over the issue of women's political participation. Women do articulate their dissatisfaction with the corrupt workings of government, particularly those led by the Congress Party since independence. For many Hindu nationalists, this corruption in contemporary India will be ameliorated in a Hindu nation guided by the morality and values of Hinduism. Renouncers, both male and female, are central to the process by which the religious politics of the nation is envisioned.

Hindu nationalist *sadhvis* draw on religious tradition to establish their authority as renouncers with the ability to penetrate worldly illusions. For those listening to their speeches, their words come with the sacred authority of a religious sermon delivered by individuals believed to have attained detachment from the world. Yet, clearly, Hindu nationalist religious exegesis cannot be separated from the political aspirations of Hindu nationalism. Talal Asad asserts that "religious symbols . . . cannot be understood independently of their relations with non-religious 'symbols' or of their articulation of social life in which work and power are always crucial" (1983:251). In other words, the meaning of religious symbols cannot be separated from the contexts in which they are deployed. Asad also argues that the symbolism of a sacred text is contingent on the workings of power through which "their correct reading is secured" (251). The *pravachans* of Hindu nationalist *sadhvis* were delivered in the context of Hindu nationalist politics and the accompanying Hindu chauvinism and xenophobia that inflects their exegesis.

I do not want to suggest that nobody in the movement questions the religious authority of the *sadhvis*. Ann Grodzins Gold and Gloria Goodwin Raheja have highlighted the importance of listening to the voices of women who challenge the dominant ideologies that inscribe their lives (Raheja and Gold 1994). Such challenges to dominant ideology, in the form of questioning renunciant authority, are evident in the Hindu nationalist movement as well. Vimla and I spent an entire afternoon irreverently gossiping about the romantic affairs of renouncers in a conversation that implied that at least some of them were frauds. Vimla told me with a laugh that there is nothing difficult in being a renouncer. She said, "You just have to look spiritual and then people will give you Mercedes cars and you will live in luxury." I mention Vimla's example here because we should not assume that all those in the movement subscribe to its ideology; instead we must pay attention, as Purnima Mankekar suggests, to the "fissures intrinsic to hegemonic discourses" (1999:255). Vimla's critique, however, was not the view that I commonly encountered during my fieldwork. Instead, for the most part, as is clear from the discussion above, *sadhvis* were talked about and treated with deference.

Conclusion

Sadhvis within the Hindu nationalist movement are deeply entrenched in the politics and ideological struggles of the world. While some might argue that they are not renouncers at all, the crucial point is that, for the most part, in practice they are viewed as renouncers and are credited with the power and the wisdom of renouncers both within and outside the movement. It is because they are renouncers that people interpret their words as "*pravachans*" that belong in the world of the sacred, are backed by spiritual authority, and are therefore qualitatively different from those delivered by Hindu nationalist ideologues. It is their ability to weave the sacred into the political realm, and to infuse religious exegesis with the political aspirations of Hindu nationalism, that makes *sadhvis* particularly effective spokespersons for the movement.

Yet, if their political power resides in their renunciant authority, their renunciant authority is contingent on their detachment from worldly involvement. Thus Hindu nationalist *sadhvis* must articulate the complementarity of the renunciant tradition with worldly involvement. While classical theories of renunciation suggest that a renouncer must be physically removed from the social world and live alone practicing austerities and gaining a higher knowledge, later religious texts like the Bhagavad Gita suggest that renunciation is compatible with worldly involvement. As I have shown, the Bhagavad Gita suggests that necessary action performed to preserve dharma and done with complete detachment from the fruits of action can in fact lead one to liberation. These ideas are central to the worldly involvement of Hindu nationalist renouncers who straddle the boundaries between the spiritual and the mundane. Not only does the Gita provide the basis for the worldly involvement of these renouncers, but it is also central to the message Hindu nationalist *sadhvis* convey to mobilize people into participating in Hindu nationalist politics. Having established India as a sacred land formed by the body of the goddess Bharat Mata, Hindu nationalist *sadhvis* use the Gita to argue that it is the sacred duty of all Hindus to participate in the struggle to establish a Hindu India.

The discursive practices of Hindu nationalist *sadhvis* are deeply gendered in multiple ways. As I have argued, *sadhvis* draw heavily on gendered imagery to motivate both men and women to embrace Hindu nationalist politics. I also suggest that it is the ambiguous position of these *sadhvis* as renouncers and as women that makes them powerful voices for defending dharma. As women who are responsible for ensuring that Hindu values and culture remain at the core of the nation and as renouncers who have a superior understanding of religious ideas, they are compelling voices to mobilize Hindus to participate in the politics of Hindu nationalism.

Acknowledgments

I would like to thank the American Institute of Indian Studies for funding this project. I am also very grateful to Ann Grodzins Gold, Meena Khandelwal, and Sondra Hausner for their comments and suggestions on earlier versions of this essay.

Notes

Epigraph. This quotation is part of a speech delivered by Sadhvi Rithambara, which I recorded in New Delhi on August 22, 1999.

1. In the speeches I recorded, Hindu nationalist female renouncers variously used dharma to refer to the moral order of the world, to duty, as well as to refer to Hindu religion. I have specified the meaning being called upon when it is not clear from the context.
2. As I wrote this article in July 2005, the senior leadership of the BJP was engaged in a public conflict over whether or not their president, L.K. Advani, should resign for departing from the RSS line (see endnote 4). Advani resigned as BJP president in December 2005.
3. Laura Ahearn convincingly argues that we must be wary of romanticizing agency as resistance. It is therefore essential to uncouple agency from progressive feminist politics (Mahmood 2004) and understand the ways in which agency often reflects and reinforces hierarchical and patriarchal power structures.
4. In June 2005, the BJP president, L.K. Advani, made several statements on a visit to Pakistan that infuriated powerful figures in the RSS and the VHP. Among other things, Advani declared that the day the Babri Masjid was destroyed was the "saddest day" in his life. Political analysts view these statements as Advani's attempt to recast himself as a moderate particularly since he was a leader of the Ramjanmabhumi movement. What the incident also reveals, however, is dissent (even if politically motivated) within the movement.

5. Khandelwal relates the story of Queen Chudala who runs a kingdom and resides amidst wealth, power, and luxury. Yet, she is so detached from these as well as from the symbols of renunciation (living alone and avoiding possessions, luxuries, etc.) that she has already attained liberation (2004:40–42).
6. It is important to note that not all renouncers would accept the view of political activity as *seva*.
7. Lise McKean (1996) has written about the Hindu nationalist construction of Bharat Mata.
8. *Sindhur* is the red powder that married women wear on their heads to symbolize their marital status. Here I have glossed *sindhuri* nights as married nights because Rithambara is referring to nights shared with their husbands as married women.
9. Sara Shniederman (this volume) also discusses the ability of female renouncers to take on male characteristics.
10. Meena Khandelwal discusses the strategic use of femininity by female renouncers she worked with (2004).
11. Islamabad, Rawalpindi, and Karachi are all cities in Pakistan. Islamabad is the capital of Pakistan.
12. Mohammad Ali Jinnah was the leader of the Muslim League and the leader of the Pakistan movement in colonial India. The movement sought to establish Pakistan as a separate state for Muslims, an idea that came to fruition on August 14, 1947.
13. Here Rithambara is referring to the conflict between India and Pakistan over the disputed territories of Kashmir.
14. The Mahabharata is an ancient Hindu epic that addresses the conflict between two royal lineages—the Pandavas and the Kauravas. Duryodhana is the eldest of the Kaurava brothers and their leader.
15. Gandhari is the wife of the blind king Dhritarashtra who is the father of the Kauravas, the cousins and enemies of the Pandavas who, at the end of the epic, after the bloody battle waged in Kurukshetra, take their place as the rulers of the kingdom.
16. Four stages in the life of all twice-born (upper-caste) men as laid out around the first century CE in the Dharmashastras include: *brahmacharya* (student), *grhasta* (householder), *vanaprastha* (forest dweller), and *sannyasa* (renouncer).
17. She used the word jungle in her speech.
18. The *sudarshan chakra* is a sacred weapon associated with Vishnu. It is used by Krishna, an incarnation of Vishnu, in the Mahabharata.
19. Krishna's sacred weapon, the *sudarshan chakra*, symbolizes the weapons that are necessary for this historical moment.
20. While I use the real names of religious renouncers in instances where they are giving public performances available to anyone, I use pseudonyms to protect the identity of all the other individuals mentioned here including the interview with a prominent female renouncer later in this chapter who I call Sadhvi Kiran Bharati.
21. These idols "miraculously" appeared inside the mosque one night in 1949 and were widely seen by the movement as a divine sign that indeed the mosque

stood on the birthplace of Ram. Anand Patwardhan, in his film, *In the Name of God* (1992), has interesting footage in which he interviews the Hindu priest attending to the idols inside the mosque who admits to having been involved in planting them there in 1949.

22. Ram uses an army of monkeys to destroy the demon king Ravana. Also, Hanuman, the monkey God who plays a critical role in the defeat of Ravana, is said to be one of Ram's greatest devotees. The suggestion here is that many associated the sight of monkeys on the roof of the mosque with a religious sign from god that indeed this was Ram's birthplace.

Bibliography

Ahearn, Laura M. 2001. Language and Agency. *Annual Review of Anthropology* 30:109–137.

Asad, Talal. 1983. Anthropological Conceptions of Religion: Reflections on Geertz. *Man* (n.s.) 18(2):237–259.

Bacchetta, Paola. 1996. Hindu Nationalist Women as Ideologues: The Sangh, the Samiti and Differential Concepts of the Hindu Nation. In *Embodied Violence: Communalising Women's Sexuality in South Asia.* K. Jayawardena and M. de Alwis, eds., pp. 126–167. New Delhi: Kali for Women.

Bacchetta, Paola, and Margaret Power. 2002. Introduction. In *Right Wing Women: From Conservatives to Extremists Around the World.* P. Bacchetta and Margaret Power, eds., pp. 1–15. New York: Routledge.

Basu, Amrita. 1995. Feminism Inverted: The Gendered Imagery and Real Women of Hindu Nationalism. In *Women and the Hindu Right: A Collection of Essays.* T. Sarkar and U. Butalia, eds., pp. 158–180. New Delhi: Kali for Women.

———. 1999. Hindu Women's Activism and the Questions it Raises. In *Resisting the Sacred and the Secular: Women's Activism and Politicized Religion in South Asia.* Patricia Jeffrey and Amrita Basu, eds., pp. 167–184. New Delhi: Kali for Women.

Bhargava, Rajeev. 2003. The Cultural Nationalism of the New Hindu. *Dissent* 50(4):11–17.

Bhatt, Chetan. 2001. *Hindu Nationalism: Origins, Ideologies and Modern Myths.* Oxford: Berg.

Blee, Kathleen M. 2002. *Inside Organized Racism: Women in the Hate Movement.* Berkeley: University of California Press.

Chatterjee, Partha. 1986. *Nationalist Thought and the Colonial World: A Derivative Discourse.* Minneapolis: University of Minnesota Press.

Elder, Joseph. 1990. Some Roots and Branches of Hindu Monasticism. In *Monastic Life in the Christian and Hindu Traditions: A Comparative Study.* Austin B. Creel and Vasudha Narayanan, eds., pp. 1–36. Lewiston: Edwin Mellen.

Flood, Gavin. 1996. *An Introduction to Hinduism.* Cambridge: Cambridge University Press.

Ghosh, Bishnupriya. 2002. Queering Hindutva: Unruly Bodies and Pleasures in Sadhavi Rithambara's Performances. In *Right Wing Women: From Conservatives to Extremists Around the World.* P. Bacchetta and M. Power, eds., pp. 259–272. New York: Routledge.

Gokhale, Nithin and Ajith Pillai. 1999. The War That Should Never Have Been. Cover Story. *Outlook Magazine*, September 6, 1999:20–27.

Jaffrelot, Christophe. 1999. *The Hindu Nationalist Movement and Indian Politics: 1925 to the 1990s.* New Delhi: Penguin Books.

Khandelwal, Meena. 2004. *Women in Ochre Robes: Gendering Hindu Renunciation.* Albany: State University of New York Press.

Lamb, Sarah. 2000. Being a Widow and Other Life Stories: The Interplay between Lives and Words. *Anthropology and Humanism* 26(1):16–34.

Mahmood, Saba. 2004. *Politics of Piety: The Islamic Revival and the Feminist Subject.* Princeton: Princeton University Press.

Mankekar, Purnima. 1999. *Screening Culture, Viewing Politics: An Ethnography of Television, Womanhood and Nation in Postcolonial India.* Durham: Duke University Press.

McKean, Lise. 1996. *Divine Enterprise: Gurus and the Hindu Nationalist Movement.* Chicago: University of Chicago Press.

Menon, Kalyani D. 2005. We Will Become Jijabai: Historical Tales of Hindu Nationalist Women in India. *Journal of Asian Studies* 64(1):103–126.

Menon, Ritu, and Kamla Bhasin. 1998. *Borders and Boundaries: Women in India's Partition.* New Brunswick: Rutgers University Press.

Miller, Barbara Stoler. 1977. *Love Song of the Dark Lord: Jayadeva's Gitagovinda.* New York: Columbia University Press.

———. 1986. *The Bhagavad Gita: Krishna's Counsel in Time of War.* New York: Columbia University Press.

Olivelle, Patrick. 1998. Mundaka Upanishad. In *Upanishads.* New York: Oxford University Press.

Patwardhan, Anand. 1992. *In the Name of God.* New York City: First Run Icarus Films.

Raheja, Gloria Goodwin, and Ann Grodzins Gold. 1994. *Listen to the Heron's Words: Reimagining Gender and Kinship in North India.* Berkeley: University of California Press.

Rajagopal, Arvind. 2001. *Politics after Television: Hindu Nationalism and the Shaping of the Public in India.* Cambridge: Cambridge University Press.

Ranganathananda, Swami. 2001. *Universal Message of the Bhagavad Gita: An Exposition of the Gita in the Light of Modern Thought and Modern Needs.* Kolkata: Advaita Ashrama.

Sarkar, Tanika. 1993. Women's Agency within Authoritarian Communalism: The Rashtrasevika Samiti and Ramjanmabhoomi. In *Hindus and Others: The Question of Identity in India Today.* Gyanendra Pandey, ed., pp. 24–45. New York: Viking.

———. 1995. Heroic Women, Mother Goddesses: Family and Organization in Hindutva Politics. In *Women and the Hindu Right: A Collection of Essays.* Tanika Sarkar and Urvashi Butalia, eds., pp.181–215. New Delhi: Kali for Women.

———. 2001. Aspects of Contemporary Hindutva Theology: The Voice of Sadhvi Rithambara. In *Hindu Wife, Hindu Nation: Community, Religion and Cultural Nationalism*, pp. 268–290. New Delhi: Permanent Black.

Sinclair-Brull, Wendy. 1997. *Female Ascetics: Hierarchy and Purity in an Indian Religious Movement*. Surrey: Curzon Press.

Snow, David A. et al. 1986. Frame Alignment Processes, Micromobilization, and Movement Participation. *American Sociological Review* 51(August):464–481.

Yuval-Davis, Nira. 1997. *Gender and Nation*. London: Sage.

Chapter 7

How Buddhist Renunciation Produces Difference

Kim Gutschow

The last time I saw Yeshe, the oldest member of Karsha's Tibetan Buddhist nunnery, she cried as I left. It was not my departure she was bemoaning, but the long Himalayan winter ahead. She had explained earlier that she was worried about finding someone to gather enough dung to heat her small cell over the next six months. Although still sprightly and fit at 76, she could no longer run up and down the steep slopes collecting yak dung like she used to. Like other elderly nuns, she relied heavily on the help of young nuns to assist her in building up a store of dried dung patties every autumn. Unlike many of the other nuns, however, she received almost no help from her immediate family in this onerous task. Abbi Yeshe had weathered many hardships in her life, but her family had caused her the greatest distress.

Yeshe was born in the year of the Tiger (1926) in Zangskar, a Himalayan region the size of Rhode Island, located in the southeastern corner of Indian Jammu and Kashmir. Raised in a middle-income household in one of Zangskar's largest and wealthiest villages, Karsha, Yeshe was courted by the son of Zangskar's most prominent official, the Zildawar, or revenue official. To her suitor's dismay, she rejected his advances and instead moved in with his aunt, Yangdzom, who lived as a renouncer on the cliff above Karsha village. Soon afterward, Yeshe and Yangdzom set off on a lengthy pilgrimage several thousand kilometers from Kashmir to Tibet, where they were both ordained as novices by one of the highest ranking monks in the land, the head of the Dalai Lama's own ruling Gelugpa sect. When they returned, the two nuns had sufficient religious clout to found a fledgling community of nuns in Karsha. One by one, other women joined them, built their own cells, and took part in the joint meditation and devotional practices. After Yangdzom passed away, Yeshe continued to organize many of the nunnery's longer-term projects such as building an assembly hall, finding the funding for an incipient Great Prayer festival and other rituals, and organizing teachers from Karsha monastery who came to instruct the nuns in basic Tantric rites and meditations.

Initially, Yeshe received plenty of assistance from her parents. Her father held a village-wide begging beer (*'dri chang*) to raise funds to build her cell.[1] Every household in Karsha was invited to send one member to an evening of beer drinking. At such events, hosts cleverly serve no food and wait for the inevitable inebriation to set in. The hosts then invite the collected guests to stand and pledge gifts to the upcoming ritual event he is stewarding. The public nature of the pledges insures a kind of competitive generosity in which guests strive to outdo their neighbors. Yeshe recalled that her kin and neighbors gave their labor and many of the supplies needed to build her cell. While her parents allowed her the use of several fields during her life at the nunnery, she fared less well after their death and that of her eldest brother.

When her brother's son inherited the family estate, he was still a young recruit in the Indian Army who only returned home a few weeks each summer while on leave. As a result, Yeshe spent her days working with his wife, a woman who never warmed to Yeshe's controlling manner. After one too many bursts of Yeshe's sharp tongue, the young wife gradually made it clear that Yeshe's help was no longer needed on the family farm. For years, Yeshe found work in exchange for food at various local houses. After decades of working tirelessly for the man who had once courted her, she had a falling out with his wife. Yeshe then found a family to till the fields her father had loaned her. In a rather unusual arrangement, Yeshe kept the grain from the field in exchange for paying him with the straw and alfalfa the field produced as well as doing some light work on the family's farm. To supplement her meager income, she asked her closest friend at Karsha's monastery, for help. Tashi, a former abbot or Zurba (*zur ba*) at the monastery, had asked for Yeshe's aid decades earlier when he had served as monastic steward. Because Tashi was too busy to care for the monastery's yaks and cows, he had asked Yeshe to spend her summers at the monastery's high pasture huts tending and milking the herd.

Zurba Tashi, like other elderly monks at Karsha's male monastery, is the frequent recipient of local gifts of butter, milk, and other supplies. He too, like Yeshe, no longer receives much help from his family. But he earns a tidy income from gifts and payments made for his ritual services. Zurba Tashi is highly versed in Tantric rituals, specifically those that propitiate the underground *klu* spirits that rule over fertility, snowfall, and water.[2] He travels across Zangskar performing rites to avert drought, cause snowfall, and restore failing springs. His propitiation of the *klu* spirits is believed to secure village and household purity and prosperity. As a result, Tashi's services are so much in demand that he can satisfy only a fraction of his ritual requests. His patrons curry his favor with spontaneous gifts throughout the year. He receives so much that he can afford to spread these gifts among his kin and friends. In contrast, and for reasons described below, Yeshe and the other nuns at Karsha earn little from ritual services because they are rarely asked to perform the pragmatic rituals to ward off disaster or disease.

Renouncing Without Rejecting Kin

This brief story implies a significant difference in the material circumstances of Zangskari nuns and monks. This essay seeks to elucidate the factors that have led villagers to call monks for ritual services and nuns for rather menial services in the village economy. While the disparity in wealth and prestige between monks and nuns has been documented, the role of kinship in sustaining this disparity has not.[3] I will explain how reciprocal relations between kin and Buddhist renouncers help shape the disparity in wealth and prestige between the female and male monastic orders. Gender discrimination is produced and reproduced through a range of reciprocal practices between Buddhist renouncers, their kin, and community. Each member in this equation performs roles complicit with the denigration of nuns. Buddhist renunciation upholds nonreciprocity as an ideal. Yet the practice of renunciation relies on obligatory exchanges and mutual services that produce a profound hierarchy between monks and nuns.

From a strictly textual basis, Buddhist renunciation appears to require a renunciation of those forms of reciprocity most shaped by kinship. The Buddhist monastic code or Vinaya specifies that ordained monks and nuns must reject sex, marriage, procreation, and individual inheritance, all of which are regulated by kinship norms in many societies. As anthropologist Stanley Tambiah (1970:68) confirms, Buddhist philosophical texts have described monkhood as "an initiation that offers a man a way out of reciprocity, a way for a man to become himself living in but not dependent upon society." Yet these textual ideals conflict with the lived reality in many Tibetan societies and the Indian Himalayas today.[4] Although many Himalayan Buddhist communities support the odd isolated meditator on an isolated retreat, most are far more engaged in daily transactions with village monks and nuns. This essay is dedicated to showing how Buddhist monks and nuns are intertwined with their local kin and communities.

In this view, Buddhist renunciation is less a rejection of kinship than its re-envisioning. Buddhist renouncers remain as closely embedded in kin relations as Hindu renouncers (Hausner and Khandelwal, this volume) or Bauls (Knight and Hanssen, this volume). Yet I will argue that there is a crucial difference in how kinship relations operate for Buddhist nuns and monks. More significantly, the difference in how nuns and monks maintain reciprocity helps explain the considerable disparity in wealth and power between the two orders.

The study of kinship has taken on new salience within gender and practice theories. Kinship is better understood as a set of strategic and contested practices and relationships than a set of algebraic rules or systematic structures determining those relationships.[5] More recent anthropological perspectives have stressed the strategic, contingent, and gendered nature of kinship systems and their reciprocities. While practice theories have highlighted the flexible and strategic nature of kinship, gender theories have sought to analyze its adaptation to sexual and cultural circumstances. Let us consider how Buddhist nuns and monks use and adapt kinship practices and terms. The result offers a modest rethinking of both Buddhist practice and kinship.

Kinship and Monasticism in Zangskar

Lying on the western edge of the Tibetan plateau, Zangskar is largely unpopulated due to its arid and inhospitable landscape of snowfields, glaciers, and steep rocky slopes. The sparse landscape supports an economy short of both food and labor. Most Zangskari households own enough land to produce a slight surplus of three staple crops—barley, wheat, and peas—as well as dairy supplies from their livestock herds. Grain, fodder, and dairy supplies are stockpiled as insurance against drought when not donated to the monastery in alms. In years of drought, the cattle may be killed for lack of fodder and loans will be taken from the monastery, the richest institution in the land. Households are nearly self-sufficient in grains although most households rely on wage labor or government jobs for supplemental incomes. The growing numbers of young men taking distant jobs in the military or government service have left many farms almost bereft of labor. As the population has expanded so has the total farmland. Yet the lack of mechanization has left households desperately searching for labor. Old people complain that "by the next generation there will be nobody but us old ones around in Zangskar" (*mtho rus sting nas nga tsho ga mo ma ni med zangs dkar nang la yod*).

An assortment of marriage patterns including patrilocal monogamy, polyandry, and polygyny are flexibly oriented toward keeping household landholdings intact.[6] Many households still practice primogeniture as the eldest son inherits the entire estate. The younger sons who are disenfranchised may cohabit with their eldest brother, settle new lands, marry out, become monks, or take up other occupations. In recent years, younger brothers are less inclined to remain dutifully in their elder brother's house and marriage. These younger sons have begun to claim their legal share of the family land, building new houses, and taking up wage jobs to supplement their dwindling farm incomes. For daughters, who are married out, the picture looks slightly different. Daughters do not ordinarily inherit any family property despite their legal right to an equal share. Customarily, brothers insist that their sisters forfeit their share of the family property upon marriage. Daughters inherit their mother's jeweled headdress (*pe rag*) and an extensive dowry from their extended kin on their wedding day. Locals deem these gifts a substitute for the daughter's share of the family land. As an exception, women can inherit the family farm if they have no brothers. Even in this case, however, the property will rarely be placed in the daughter's name, but simply be transferred directly from her father to her first son. In sum, all daughters are effectively disinherited while younger sons are no longer quite as disenfranchised as they once were.

How do these kinship practices relate to the motivations for renunciation? Eldest daughters and sons rarely become monks or nuns because they receive the family estate or jewels. Indeed, not one of the 140 nuns and monks I interviewed in Zangskar in the last 15 years was an eldest daughter or son from a legitimate marriage. Most nuns and monks in Zangskar are younger siblings unless they are stepchildren, children of divorce, or born out of wedlock. Ruptures in kin relations such as being orphaned, adopted, or born from an illegitimate union quite frequently destine a child for the monastery or nunnery.[7] However, such difficult circumstances hardly sustain a child's entrance and career in the monastic order. Women do not arrive at the nunnery gate "by accident" because of unfortunate childhoods or marital circumstances. Instead, most young girls who remain in the monastic order must show significant religious devotion or aptitude during their apprentice years. The lengthy apprenticeship under the surveillance of the elder nuns turns many prospective applicants away long before novice ordination.

Overall, the number of monks appears to have leveled off due to changing economic conditions and shifting inheritance patterns. Younger sons have far less incentive to become monks now that they can claim their share of the family land and pursue salaried jobs or wage labor in the growing cash economy. Additionally, new forms of wealth and prestige have undermined the preeminence of monks. For the younger generation, the worldly swagger of money and military uniforms are prized far more highly than meditative prowess. Yet daughters have not been able to claim their rightful share of family property, nor have they been as successful in obtaining jobs in the civil service and military. Outside of the fields of education and medicine, women have faced stiff competition for government jobs given the rapidly expanding number of candidates and only slight increases in available jobs. As a result, the nunnery remains a desirable career option for women who wish to pursue education, service, and travel. The rising number of applicants at existing nunneries and construction of many new nunneries in the region in the 1990s implies the rising demand for renunciation among women. In 1998, five percent of Zangskar's 12,000 inhabitants were monks and nuns, mostly of the Gelugpa and Drugpa Kadyud schools of Tibetan Buddhism.[8]

The monastery still offers far more opportunity for social mobility than the nunnery does. Those monks from higher social and economic ranks can rely on family status to facilitate their rise through the monastic bureaucracy. Alternatively, for those monks from humble origins, the monastic profession offers travel and educational opportunities unavailable to most laymen. Today as in the past, Zangskar's poorest monks have benefited from reserved seats at the great Gelugpa monastic colleges. Before 1959, Zangskari monks made the arduous journey to Tibet to continue their studies. Nowadays, young monks take buses and trains to Karnataka State, where the Gelugpa colleges have relocated in exile. The exposure to India's cosmopolitan culture, popular media, and foreign tourists lures some monks out of robes and makes many more sophisticated purveyors of their religion. Those monks who advance furthest in the transnational religious scene may find themselves making mandalas for audiences in far-flung cities like Sydney and Boston. These opportunities are rare for nuns. First, there are no reserved seats for Himalayan nuns at Tibetan nunneries. Second, most Himalayan nuns do not have as extensive a network of foreign sponsorship as monks do.[9] While nuns may go on pilgrimage to Dharmsala, Nepal, and South India, only the luckiest manage to remain for purposes of study. Even then, only a few select nunneries teach the arts of mandala making, the most likely vehicle for sending nuns abroad.

If You Want a Servant, Make Your Daughter a Nun

Despite the rapidly changing economic and social climate of Ladakh and Zangskar, nuns and monks still rely on their kin in many traditional ways.

While nuns often *ask* their families for money, monks are more likely to *offer* their families financial help. As a result, most families view a daughter becoming a nun somewhat differently than they do a son becoming a monk. A Tibetan proverb explains: "If you want a servant, make your daughter a nun. If you want to serve, make your son a monk." When parents send a child to the nunnery, they earn her labor for the foreseeable future. When they send a son to the monastery, they will serve his needs in exchange for ample payback from his ritual and religious career. Sending a son to the monastery is like enrolling him in an Ivy League or Oxbridge University on a full scholarship. His elite education guarantees ample opportunities for privilege and profit for the rest of his life. By contrast, sending a daughter to the nunnery is like sending her to a state university, without scholarship. She must pay her own way, but she will remain near enough to be of use on the family farm. As a result, nuns are both a burden and a benefit to their natal household. They require a share of the household's limited food supply even as they offer assistance through their frequent labor contributions.

Nuns use and are used by their kin. Customarily, nuns are expected to contribute their share of labor to the family estate long after they join the assembly of nuns. Those families who send a daughter to the nunnery earn a lifelong adult laborer without children or a husband. Nuns can strategize to work as wage laborers, but only if there is no pressing need at home. During the busiest agrarian seasons—plowing, first watering (*sgrol chu*), and harvest—nuns work for only little compensation. Other villagers can sell their labor to the highest bidder and look for work elsewhere if the wages or treatment are undesirable. But nuns may hesitate to offend the families to whom they are beholden. However, nuns are not unaware of their predicament. I asked my roommate, Garkyid, how she could sit cross-legged all day, when my knees were aching. She quipped, "How can I complain! The Buddha has given us a holiday [from chores at home]." Some of the younger nuns at Karsha marshaled a subtle transcript of resistance when they refused to make breads for local wedding feasts one year. They decided that laywomen were as capable and more suitable for this onerous chore. After their Tibetan philosophy teacher lobbied on their behalf, most of the younger nuns stopped going to bake breads. Yet a few elderly nuns continued this customary duty, noting that they could not abandon villagers who had helped them so often in times of need.

Like most nuns Garkyid spends many of her days at her natal household, working for her elderly mother, a widower. She walks to her mother's house several times a week to help her mother work in the fields, feed or milk the livestock, wash, roast, and grind the grains, churn butter, collect dung, spin wool and make clothes or boots, bake breads, or make barley beer for upcoming village and household festivals. In exchange, she returns each evening to the nunnery bearing a bit of roasted barley flour, butter, milk, or yogurt. She explained, "Because my father and eldest brother are dead and my [other] brother has only small children, my mother needs my help. I can ask my mother for grain, flour, milk, butter, meat, and yogurt, but I'd be ashamed to ask others for such food." While there is no shame in asking kin for food, it is considered impolite to ask strangers for food in Zangskar. Garkyid cannot ask her second brother for much help because he has no salary and can barely feed his own large family. When not working on her mother's fields, Garkyid works for her cousin, who earns a comfortable government salary, or as a wage laborer on local fields.

While monks are largely exempt from household labor exchanges, it is common for nuns to represent their family in village work projects—repairing irrigation channels, shepherding animals, or planting willows. Those nuns who live alone with elderly parents or single mothers are most affected by the chronic shortage of labor. One of the nuns at Karsha, Tsomo, spent most of her days taking care of her aged mother and her mother's ailing sister, who was also a nun. Although Tsomo and her aunt had a spacious cell at the nunnery, they hardly had a chance to enjoy it, as they spent their days in the cramped house with Tsomo's mother. During her mother's turn to watch the neighborhood flock, Tsomo was the one called to race after the nimble goats and sheep. Typically, up to one-third of the nun's assembly is absent from the trimonthly ritual ceremonies due to labor obligations elsewhere. During the harvest season, the nunnery may be abandoned by all but the weakest or most elderly nuns.

Nuns are in a double bind. If they ignore their family demands, they lose the means of subsistence. Yet they also incur outrage if they neglect their duties to the community of nuns. An inescapable cycle of expectation and guilt binds nuns to their families and their assembly. One spring morning just after dawn, Garkyid's elderly mother paid her a rare visit at her cell. The nuns had been out collecting limestone for the annual plastering and whitewashing of the assembly hall. Garkyid was rather surprised to see her mother at the nunnery, whose arthritis had kept her away for months. Her mother was desperate for Garkyid to come home that minute to read the sacred texts at a village-wide ritual her household was sponsoring. Garkyid flatly refused because she did not want to forfeit her obligations at the nunnery. Yet she promised to help her mother and dashed off to the monastery to find a monk to do the reading. The next day, Garkyid arose in the predawn darkness to check that the ritual reading of the text had been completed. By daybreak, she'd returned to finish helping the nuns whitewash the hall.

In exchange for her services, a nun can expect to receive the basic necessities of food and shelter, even in times of shortage. She may be given stockpiles of grain and other foods as well as clothing, utensils, bedding, furnishings, and cash. Either the nun or her family can terminate these reciprocal relations due to disputes, albeit not without cost. Because nuns spend most of their days working for their families, they tend to join a nunnery located within walking distance of their home. Those few families that do send a daughter to a more distant nunnery usually find a nearby household to "adopt" their daughter in exchange for her labor. This adoptive relationship relies on a fictive or distant kinship, in which the nun pledges her labor and loyalty in exchange for material support. The nun will address and be addressed by the members of this household as if they were her kin.

In some ways, nuns are like unmarried daughters who reside in their natal households until death. Both nuns and unmarried daughters provide valuable labor services to their families in exchange for daily economic support, unlike daughters who get married and rely on their in-laws. Yet while aged spinsters reside in their natal household either with their elder brother or their aged parents, nuns do not remain at home permanently, nor do they rely solely on their family for material support. After joining the assembly, most nuns build, buy, or borrow a stone cell at the nunnery where they will live until death. The importance of this shift in residence is marked by the phrase for joining the monastic assembly, namely "dwelling on the cliff" (*ri la bzhugs byes*). Although nuns do gain some agency by having this "room of their own," their autonomy is constrained by social and economic pressures. Nuns have as much difficulty in earning incomes as the Victorian women described by Virginia Woolf (1929). Their meager income from ritual duties may be supplemented by household and wage labor. Yet they cannot neglect their family's and society's expectations without considerable cost.

The nuns' assembly functions like an extended family bound by a fictive kinship. Younger nuns who answer to niece or granddaughter (*tsa mo*) call the elder nuns grandmother (*a phyi*) or auntie (*a ni*). Agemates or peers call each other older sister (*a che*) or younger sister (*no mo*), depending on age. The affection among the nuns transcends kinship and customary ties among agemates in the village. I have spoken to many laywomen who envy the deep solidarity that nuns build in the company of women. Close friendships are sanctioned, but sexual intimacies are avoided. While I have been privy to a host of village scandals including abortions and rapes, I have yet to hear of lesbian relations among nuns. Although local banter suggests that monks do sneak into village bedrooms, there is less talk of sexually frustrated nuns.

A tacit and mutual web of reciprocity rules over the community, as nuns are expected to help each other in numerous ways. The elder nuns train and tutor the younger nuns during the apprenticeship period in ritual duties as well as mundane chores like cooking and serving tea for the assembly.

In exchange, younger nuns serve the elders long after their apprenticeship is complete by fetching water and dung during the winter months and performing much of the heavier labor for the community. It is common for apprentice nuns to live with their tutors. Yet joint living arrangements between unrelated nuns are discouraged after the apprenticeship is complete. Relatives like sisters or aunts and nieces usually cohabit and work together, but even unrelated nuns are obligated to help one another as if they were family. Most nunneries have a few peer groups of nuns bound by age and friendship. These groups require loyalty and mutual assistance. Peers often exchange days on each other's family fields, while they eat together in the mornings and nights.

As nuns age and their family ties dissolve, they rely on their own monastic community more and more. Women begin and end their life at the nunnery in an infant-like dependence on others. When Yeshe recently fell dreadfully ill with hepatitis, she was forced to rely on Zurba Tashi and other nuns rather than her immediate family, who had rejected her. During her ailment, her nephew's wife and family never paid a visit although Tashi and other villagers offered their consolation. Thuje, a younger nun who had studied Tibetan medicine with her father and trained as a community health worker, nursed Yeshe back to health all winter. Yet Thuje was far more than a nurse as she spent the winter gathering dung, cooking, and fetching water for Yeshe. In gratitude, Yeshe spent months during her recovery making Thuje a woolen jacket and boots.

For ordained nuns, belonging to a monastic community can mean the difference between mild and severe destitution. A friend once took me to meet an elderly, unordained renouncer who lived in a cave above Padum. She was too weak to look for dung or beg for food. In a voice faint with despair and malnutrition, she told me that she had eaten only barley gruel that week. I saw little sign of food in her cave, only a few broken kettles, and a single stone pot sitting on a cold hearth with no firewood, kerosene, or matches in evidence. It was not clear how her situation would improve as her immediate family had long since passed away. While we brought her supplies that day, my friend soon organized the local woman's group to provide her with a more steady source of rations.

The Economics of Renunciation for Nuns and Monks

Nuns depend on their kin for daily food because their institutions do not have adequate economic resources. By contrast, monks can rely on their monastery's vast endowments of land, livestock, and capital. Monks also receive more alms, ritual donations, and ritual service payments than nuns do. In what I've called an "economy of merit," both social and symbolic capital concentrate among monks and monasteries (Gutschow 2004).

Historically, permanent endowments of land and livestock were given predominantly to monasteries and prominent monks. By the turn of the twentieth century, monasteries were the wealthiest landholders in Zangskar. While monasteries held one-tenth of all cultivated land in Zangskar, most nunneries owned no land or livestock. In 1994, Zangskar's largest monastery, in Karsha, owned 90 times as much land and ten times as many yaks and cattle as the average household. Yet its largest nunnery, also in Karsha, owned less than half an acre.[10] Karsha's monastery receives more than one hundred times as much grain as its nunnery does annually from alms and sharecropper rents. Yet there are only four times as many monks as nuns in Karsha.[11] Roughly half of the 11,000 kilos of grain collected annually at Karsha goes directly to the monks. The other half is stored for ritual services, loans, or other institutional needs. While Karsha's nuns receive a few kilograms annually from their institution, the monks each receive more than 50 kilograms of grain annually. When laypeople make a donation to a monastic assembly, each monk or nun receives a share (*skal ba*) of this offering.[12] Individual monks in Karsha may collect up to several thousand rupees in earnings, while nuns earn no more than a few hundred rupees.

The monastery is as much a place of economic and social mediation as of symbolic meditation. Although monks do find time to meditate later in life, their early years are spent managing the monastery's assets, debits, and investments and ritual bureaucracy. From the moment they join the assembly, monks rotate through the monastic offices as cooks, tea bearers, trumpeters, ritual assistants, and stewards (*gnyer pa*). Those more textually inclined monks may take up more challenging roles as disciplinarian (*dge skos*), chantmaster (*dbu mdzad*), ritual master (*rdo rje slob dpon*), and abbot (*mkhan po*).[13] By comparison the retinue of posts at the nunnery is rather circumspect and all nuns must rotate through every post, from sacristan all the way up to chantmaster, who doubles as head of the nunnery. Zangskari nunneries do not have female abbots. They are ruled by male abbots who usually rule in absentia. During disputes or in cases of misconduct, the male abbot has final authority to discipline or expel a nun from the assembly.

Karsha monastery draws its members from a dominion (*mnga' 'og*) that spans much of northern and central Zangskar. This domain includes those villages that rely on the monastery for ritual services. By contrast, Karsha and most other nunneries draw most of their members from one or two villages to which the nuns can walk for their daily work. This turns out to have profound consequences for kinds of interaction monks and nuns have with their kin.

The monks who serve three-year terms as stewards, managing the monastic treasury and its permanent endowments, mobilize extensive kin networks for assistance. They may call on their kin to help arrange begging beers and fund-raising feasts that enable them to solicit donations and obligatory payments of grain, butter, and firewood for the largest and most spectacular festivals. In exchange, kin who travel from across the region to attend and donate gifts to these festivals expect to be hosted by their relatives at the monastery. On the flip side, kin will host monks during the collection of alms, sharecropper rents, interest payments, or loans that are past due, as well as aid monks in negotiating water rights or purchasing properties. These same kin benefit from their access to the monks in charge of the monastery's most lucrative business and construction ventures. I have heard of several monks admonished but not disrobed for channeling thousands of rupees to their own families. The tacit social acceptance of this corruption underscores the necessary and symbiotic relationship between the monastery and its lay clients.

By contrast, the nuns' more reduced ritual and economic role in village life limits the kinds of assistance they can offer or require from their kin. Lacking land and other endowments, the nunnery relies on its members to fund many rituals. The regular trimonthly rituals at the nunnery are stewarded by nuns, on rotation, who collect the necessary barley, butter, and other ritual foodstuffs from their families or local villagers.[14] The Great Prayer Festival and the annual fire sacrifice rely partly on a small endowment of goats that are divided among the member nuns. During these rites, each nun brings in an annual payment of butter and flour as a symbolic "interest" on the goat she has been loaned from the nunnery's herd. Yet the paltry sums collected pales in comparison with the vast stockpiles the monastery gathers from its sharecroppers. The generosity of villagers often dictates the extent of the ritual calendar at many nunneries. Many of Zangskar's poorest nunneries sharply curtail their rites for lack of endowments.[15] While nuns in Karsha are responsible for collecting an annual quota of dung and firewood for the nunnery hearth, the monastery can rely on its clients and sharecroppers for such services. I recall the scandal caused when one set of villagers refused to deliver their firewood quota because of a grazing dispute. When the monks threatened to stop performing funerary rituals for the boycotters, the villagers hastily agreed to deliver the firewood. The nunnery does not have this kind of bargaining power because its members are not called to perform the most critical life-cycle and village rites.

Unlike nuns who are rarely called to perform more pragmatic Tantric rites, monks also earn considerable incomes performing household and village rites.[16] For much of their tenure at the monastery, most monks are stationed for three years on rotation in villages across the monastic domain. When serving as ritual officiants (*mchod gnas*), monks are responsible for performing the birth, death, and other purificatory or expiatory rites that ward off sickness, misfortune, and other impurities. It is monks, not nuns, who perform rites to expel demons, ghosts, and other misfortune through the tossing, burning, and burying of effigies, threadcrosses, and ransom figures (*be le 'phang byes brgya bzhi, 'chi gnon, rgyal mdos, dgu mig bzlod*). Monks are also more likely than nuns to be called for the seasonal or ad hoc rites that ensure prosperity, fertility, and purity, while warding off slander, envy, or other negativity. Additionally, villagers tend to conscript monks rather than nuns to ward off natural disasters, and bless human ventures like journeys, constructions, and seasonal activities that meddle with the earth (plowing, watering, threshing).

Because laypeople think that monks have more merit and purity than nuns do, villagers prefer monks for most ritual expiations, propitiations, and ablutions. In Zangskar, ritual efficacy is rather like medical efficacy: the more expert the practitioner, the more effective the rite is considered to be. Because the demand for the most qualified Tantric experts exceeds their supply, kin loyalty may be called upon to obtain a desired monk's services. Families frequently request their own relatives at the monastery to arrange the performance of these rites. Nuns can and do participate in rituals like weddings, funerals, and certain village-wide rites such as honoring the ancestors (*dge tsha*) or circumambulation of the fields (*'bum skor*). However, while both male and female assemblies are called to generate merit, only monks perform the complex Tantric mediations that transfer merit, capture wealth, defeat enemies, or release the consciousness from the body during these ritual occasions. Laypeople do ask nuns to perform basic rites to generate merit for all sentient beings, including the donor. However, if more pragmatic ritual effects are desired, monks are often preferred.

Conclusion

Nuns and monks are bound to laypeople by a variety of reciprocal obligations. Yet each order differs in the types of services they provide. Nuns perform the most menial labor while monks serve as ritual and scholastic experts for their communities. While nuns may renounce marriage and procreation, they cannot avoid being called into the spheres of production and consumption by village households. Monks do work in fields on occasion; Yet it is nuns who are expected to perform daily household chores. As a result, nuns may earn material resources but little other social capital. Monks, by contrast, earn and offer highly convertible forms of economic, social, and symbolic capital. Monks provide their kin with material resources and status through their access to the largest landholder in Zangskar. The social networks and fame they earn in the course of their ritual and financial duties on behalf of the monastery generates considerable social capital. Finally, monks earn considerable symbolic capital through their ritual performances.

> To understand ritual practice, to give it back both its reason and its raison d'etre without converting it into a logical construction or a spiritual exercise, means more than simply reconstituting its internal logic. It also means restoring its practical necessity by relating it to the real conditions of its genesis, that is, the conditions in which both the functions it fulfils and the means it uses to achieve them are defined. (Bourdieu 1990:97)

Bourdieu's focus on the practical over the rhetorical logic of ritual activity led him to define symbolic capital as misrecognized capital because it appears to have been produced by disinterested means. The merit or purity produced in the course of Tantric rites offers a prime instance of symbolic capital because it appears to have been generated in a disinterested fashion. Doctrinally, monks generate merit disinterestedly for the benefit of all sentient beings. From an analytic perspective, however, monks enact an interested relationship when they perform ritual mediations whose effects are invisible if not unknowable in exchange for visible and calculable payments of cash and grain.

Monastic rituals reify existing social hierarchies as well as legitimate the accumulation of economic and social capital among monks. Buddhist doctrine claims that nobody can purify another. Yet Zangskari monks exercise and earn ritual and economic power as they purify bodies and spaces and generate merit for their patrons. The classic symbols of Tibetan Buddhism—purity/pollution, merit/demerit—may be arbitrary, yet their effect is not. Moreover, it is the ritual use of these symbols that generates power and hierarchy. Monastic rituals authorize and reinscribe an ideology that claims monks to be more pure and meritorious than nuns. The ideology of purity and pollution that ranks male over female and aristocrat over outcaste in Zangskar is reinscribed and validated by Buddhist rituals. Transient impurities—from accidental contact with outcastes or demonic sprites—can be remedied through monastic ritual. Yet lasting forms of impurity—the female body, being born into a lower caste—can only be undone in a future rebirth by making merit in this life. As a result, even the lowest positions in the social hierarchy are conscripted by the ideology of purity and pollution. In short, villagers and nuns are complicit in their own subordination when they call monks to generate merit or purity on their behalf.

Although nuns do not become submissive wives, they remain dutiful daughters and sisters. In Zangskar, married daughters are exchanged for milkprice (*o ma'i rin*)—which is given to the mother as payment for nursing the future bride, brideprice, and the promise of future reciprocities with the in-laws. By contrast, nuns are exchanged for merit and the promise of labor. Nuns may elude the patriarchal economy of marriage, but they cannot escape the exchange of women. The anthropologist Levi-Strauss argued that the universal exchange of women marked the boundary between nature and culture, separating humans from animals.[17] His theory has been roundly criticized for positing women as objects but never subjects of this exchange, and for its assumption of universality. Yet, his point that society is founded upon the exchange of women is relevant.

We must view monasticism not so much as a rejection of kinship as an alternative social relationship. Gayle Rubin (1975:177) defines "exchange of women" to mean that "men have certain rights in their female kin, and that women do not have the same rights in their male kin." This, in fact, is what happens in the monastic realm. Monks, as abbots, have rights to discipline or expel nuns, which nuns never have at the monastery. These rights give monks a certain authority over the traffic in women into and out of the nunnery, just as fathers manage the traffic of brides between households (see Gutschow 2001).

Buddhist renunciation is often portrayed as more egalitarian than its Hindu counterparts. However, both Buddhist and Hindu renunciation remain bound by prevailing social hierarchies. While Buddhist doctrine offers the appealing argument that gender has no bearing on the potential for enlightenment, its monastic practices both legitimate and extend gender disparities. In closing, Buddhist monasticism has proved somewhat less enlightened than expected.

Acknowledgments

I'd like to thank audiences at the Laboratoire d'Ethnologie et de Sociologie Comparative at the University of Paris, the South Asia Center at the University of Wisconsin in Madison, and the Judy Collins Smith Art Museum at Auburn University for their input on previous versions of this paper. Thanks to Meena Khandelwal, Sondra Hausner, and Ann Grodzins Gold for their keen editorial insight as well as to Sarah McClintock, Sarah Levine, and Isabelle Riaboff for helpful conversations on ideas developed in this paper.

Notes

1. Gutschow (2004) describes the myriad uses of Zangskar's most common religious fund-raising vehicle, the begging beer.
2. Tibetan Buddhism relies heavily on Tantric meditations and ritual performances, which have both philosophical and pragmatic aspects. Although ultimately, Tantric practices are dedicated to help individuals achieve enlightenment through the transcendence of mundane reality and duality, they also have a wide range of practical uses in village religion including the preservation of health and wealth within the community. Samuel (1993) offers an excellent analysis of the tension between the clerical and shamanic aspects of Tibetan Buddhism, while Gutschow (2004); Crook and Osmaston (1994); and Riaboff (1997) analyze Tibetan Buddhist monasticism in Zangskar from both pragmatic and Tantric perspectives.
3. See Gutschow (2004), Arai (1999); and Grimshaw (1983, 1992); on the material circumstances of Buddhist nuns compared with its monks and Khandelwal (1997) on how female Hindu renunciants differ from their male counterparts.
4. The Indian Himalayan regions of Ladakh and Zangskar have proved excellent locations to study the social and economic basis of Tibetan Buddhist monasticism (see Gutschow 1998, 2004; Riaboff 1997; Crook and Osmaston 1994; Tsarong 1987; and Grimshaw 1983). For historic Tibet, Carrasco (1959:124) describes that while monks earned a share of the income of the monastery, hostel (*khams tshan*), or college (*grva tshang*) to which they belonged, this income was usually supplemented by economic activity and family support. Carrasco notes that the economic relations of support that families provided to monks is best documented for Spiti, although he fails to mention the data for neighboring Ladakh and Zangskar.
5. Dumont (1980) and Levi-Strauss (1969) epitomize the more structuralist approaches to kinship and exchange, while Pierre Bourdieu (1977), Sherry Ortner (1996), Susan McKinnon (1991), and Mary Steedly (1992) have described kinship in terms of strategic, fluid, and gendered practices.
6. Goldstein (1971) describes the monomarital principle as a practice whereby there was only one marriage in each generation in order to avoid splitting the family property. Zangskari marriage and residence patterns are analyzed in Gutschow (1995, 2004) and Crook and Osmaston (1994). Compare Childs (2004), who describes demographic patterns in Nubri, and Aziz's (1978) account of marriage and residence patterns in historic Dingri.
7. Compare Shneiderman (this volume), who notes that in Lubra, all families with three or more daughters made at least one a nun. In correspondence (June 13, 2005), she added that while most families chose the second daughter to be a nun (rather than later daughters) she was unable to find a consistent answer as to why this was so.
8. Gutschow (2004:274) reports that Zangskar's ten nunneries housed a 125 nuns while its eight monasteries housed over 300 monks. In 2003, there were roughly 85 monks and 23 nuns in Karsha. Although 95% of Zangskar's inhabitants are

Tibetan Buddhists, there is a small minority of Sunni Muslims who have settled mostly in the central valley around Padum, the administrative center. These families are descendants of Muslims who came in the wake of the Dogra conquest in the nineteenth century and, more recently, with the influx of government servants from the Kashmir valley.

9. Several of the most avante garde nunneries in Kathmandu have begun to teach mandala making and have sent their nuns abroad, as described by Kerin (2004). Gutschow's (2004) last chapter details the effects of the rising tide of international aid at Zangskar's and Ladakh's nunneries. While this aid has contributed to substantial improvements in the economic conditions of nuns, it has proved much harder to change social attitudes toward nuns in the region.
10. Gutschow (1998) and Riaboff (1997) have discussed the statistics on land ownership in Zangskar.
11. If the membership were proportional with its sharecropping income, the monastery should have 2,000 monks. Karsha nuns receive 2.5 kg of grain annually, while Karsha monks receive roughly 60 kg of grain per year. By comparison, a Sherpa nunnery in Nepal provided each of its 23 members with 84 kg of grain per year, as Fürer-Haimendorf (1976) and Aziz (1976) noted. French (1995:4) reported that monks in one Tibetan monastery received 110 kg of grain annually before 1959.
12. When grain or monetary gifts given to the entire assembly is distributed, each monk receives a single share, and additional shares are distributed according to office. The titular head of the monastery, Ngari Rinpoche, gets five shares; the Labrang treasury gets ten shares; the abbot gets three shares; the ex-abbots, Vajra master, chant master, assistant chant master, disciplinarian, Labrang managers, and Labrang assistant each get an additional share.
13. The inner retinue (*nang 'khor*) includes the posts of: disciplinarian (*dge skos*), assistant disciplinarian (*dge g.yog*), sacristan (*dkon gnyer*), butcher (*bsha' pa*), cook (*byan ma*), and hearth assistant (*thab g.yog*). The outer retinue (*phyi 'khor*) includes: two conch blowers (*dung pa*), two long horn blowers (*dung chen pa*), and two oboe players (*rgya gling pa*). The sacristan performs offerings to the Tantric deities and local protectors.
14. Each of these monthly rituals requires the following: 1.3 kg of butter for tea and butter lamps, 7 kg of roasted barley flour for the communal offering cakes (*tshogs*), 10 kg of wheat flour for the breads (except on the tenth of every month), 1 bottle of buttermilk as leavening agent, a handful of salt, 2 handfuls of loose green tea, and a plateful of sweets known as *tshogs zas*.
15. Tungri, Dorje Dzong, and Pishu nunnery host bimonthly prayer sessions and the Great Prayer Festival, while the newest and least endowed nunneries like Skyagam, Bya, Manda, and Sani may only celebrate a brief Prayer Festival (*smon lam*).
16. Gutschow (2004); Crook and Osmaston (1994); and Riaboff (1997) describe the kinds of Tantric rites that monks generally perform from the monastic perspective, while Phylactou (1989) and Day (1989) analyze the purpose of these rites from the lay or household perspective. Although Tantra is dedicated to transcending duality in order to bring the practioner closer to enlightenment, it has many pragmatic ritual applications in Tibetan Buddhist culture.

17. Levi-Strauss (1969) believed that the universal prohibition of incest was less a rule preventing marriage within a group than one obligating groups to seek women from other unrelated groups or clans, thereby initiating the exchange of women. Books or essays dedicated to rebuking his theory on the exchange of women include Strathern (1988), MacCormack and Strathern (1980), and Rubin (1975) among others.

Bibliography

Arai, Paula. 1999. *Women Living Zen: Japanese Soto Buddhist Nuns*. Oxford: Oxford University Press.

Aziz, Barbara. 1976. Views From the Monastery Kitchen. *Kailash* 4(2):155–167.

———. 1978. *Tibetan Frontier Families*. New Delhi: Vikas Publishing House.

Bourdieu, Pierre. 1977. *Outline of a Theory of Practice*. Richard Nice, trans. Cambridge: Cambridge University Press.

———. 1990. *The Logic of Practice*. Richard Nice, trans. Stanford: Stanford University Press.

Carrasco, Pedro. 1959. *Land and Polity in Tibet*. Seattle: University of Washington Press.

Childs, Geoff. 2004. *Tibetan Diary: From Birth to Death and Beyond in a Himalayan Valley of Nepal*. Berkeley: University of California Press.

Crook, John, and Henry Osmaston. 1994. *Himalayan Buddhist Villages*. New Delhi: Motilal Banarsidass.

Day, Sophie. 1989. Embodying Spirits: Village Oracles and Possession Ritual in Ladakh, North India. PhD dissertation, Department of Anthropology, London School of Economics.

Dumont, Louis. 1980. *Homo Hierarchicus*. Chicago: University of Chicago Press.

French, Rebecca. 1995. *The Golden Yoke: The Legal Cosmology of Buddhist Tibet* Ithaca, NY: Cornell University Press.

von Fürer-Haimendorf, Christopher. 1976. A Nunnery in Nepal. *Kailash* 4(2): 121–154.

Goldstein, Melvyn. 1971. Stratification, Polyandry, and Family Structure in Central Tibet. *Southwestern Journal of Anthropology* 27(1):64–74.

Grimshaw, Anna. 1983. Rizong: A Monastic Community in Ladakh. PhD dissertation, Department of Anthropology, Cambridge University.

———. 1992. *Servants of the Buddha: Winter in a Himalayan Convent*. London: Open Letters Press.

Gutschow, Kim. 1995. Kinship in Zangskar: Idiom and Practice. In *Recent Research in Ladakh 4 & 5: Proceedings of the Fourth and Fifth International Colloquia on Ladakh*. Henry Osmaston and Phillip Denwood, eds., pp. 334–346. London: School of Oriental and African Studies.

———. 1998. An Economy of Merit: Women and Buddhist Monasticism in Zangskar, Northwest India. PhD dissertation, Department of Anthropology, Harvard University.

———. 2001. Women Who Refuse to be Exchanged: Nuns in Zangskar, Northwest India. In *Celibacy, Culture, and Society: The Anthropology of Sexual Abstinece*. Elisa J. Sobo and Sandra Bell, eds., pp. 47–64. Madison: University of Wisconsin Press.

———. 2004. *Being a Buddhist Nun: The Struggle for Enlightenment in the Himalayas*. Cambridge, MA: Harvard University Press.

Kerin, Melissa. 2004. From Periphery to Center: Tibetan Women's Journey to Sacred Artistry. In *Women's Buddhism, Buddhism's Women: Tradition, Revision, Renewal*. Ellison Findly, ed., pp. 319–338. Boston: Beacon Press.

Khandelwal, Meena. 1997. Ungendered Atma, Masculine Virity, and Feminine Compassion. *Contributions to Indian Sociology* (n.s.) 31(1):79–107.

Levi-Strauss, Claude. 1969. *The Elementary Structures of Kinship*. Boston: Beacon Press.

MacCormack, Carol, and Marilyn Strathern, eds. 1980. *Nature, Culture, and Gender*. Cambridge: Cambridge University Press.

McKinnon, Susan. 1989. *From a Shattered Sun: Hierarchy, Gender, and Alliance on the Tanimbar Islands*. Madison: University of Wisconsin Press.

Ortner, Sherry. 1978. *Sherpas Through Their Rituals*. Cambridge: Cambridge University Press.

———. 1989. *High Religion: A Cultural and Political History of Sherpa Buddhism*. Princeton: Princeton University Press.

———. 1996. *Making Gender: The Politics and Erotics of Culture*. Boston: Beacon Press.

Phylactou, Maria. 1989. Household Organisation and Marriage in Ladakh-Indian Himalaya. PhD dissertation, Department of Anthropology, London School of Economics.

Riaboff, Isabelle. 1997. Le Roi et le Moine: Figures et principes de pouvoir et de sa légitimation au Zanskar (Himalaya Occidental). PhD thesis, Laboratoire d'Ethnologie et de Sociologie Comparative, Université de Paris X.

Rubin, Gayle. 1975. The Traffic in Women: Notes on the "Political Economy" of Sex. In *Toward an Anthropology of Women*. Gayle Rubin, ed. New York: Monthly Review Press.

Samuel, Geoffrey. 1982. Tibet as a Stateless Society and Some Islamic Parallels. *Journal of Asian Studies* 41(2):215–229.

———. 1993. *Civilized Shamans: Buddhism in Tibetan Societies*. Washington, DC: Smithsonian Press.

Steedly, Mary. 1993. *Hanging Without a Rope: Narrative Experience in Colonial and Post-Colonial Karoland*. Princeton: Princeton University Press.

Strathern, Marilyn. 1988. *The Gender of the Gift*. Berkeley: University of California Press.

Tambiah, Stanley. 1970. *Buddhism and the Spirit Cults of Northeast Thailand*. Cambridge: Cambridge University Press.

Tsarong, Paljor. 1987. Economy and Ideology on a Tibetan Monastic Estate in Ladakh: Processes of Production, Reproduction, and Transformation. PhD dissertation, Department of Anthropology, University of Wisconsin at Madison.

Woolf, Virginia. 1929. *A Room of One's Own*. London: Harcourt Brace Jovanovich Publishers.

Chapter 8

Renouncing Expectations: Single Baul Women Renouncers and the Value of Being a Wife

Lisa I. Knight

Introduction

Baul renouncers of West Bengal and Bangladesh pose a number of challenges to South Asian paradigms of renunciation where one expects to see an individual severing ties for a life characterized by celibacy, itinerancy, and worldly detachments. Bauls follow none of these accepted notions of renunciation, yet they use terms like *sannyasi, tyagi*, and *bairagi* to describe themselves as renouncers.[1] In stark contrast to more standard expectations of celibacy, the hallmark of Baul renunciation is ritual sexual practices, and Bauls who become renouncers are expected to do so as a couple. Furthermore, renunciation does not necessarily indicate a dramatic transformation from householder status to solitary ascetic, as is encouraged by many other renunciant traditions, and Bauls frequently maintain their ties with family and community. Although many Bauls travel to perform Baul songs or to beg for alms, their lives are not necessarily defined by itinerancy as many remain householders or reside long term in ashrams. Further confounding the meaning of this act is the fact that renunciation is not upheld as a necessary step on the Baul path at all; Bauls maintain they can reach the same goals whether remaining within or stepping outside householder roles. Yet despite these ambiguities around the practices and expectations of Baul renunciation, it is not uncommon for

Bauls to take a formal rite of renunciation with a guru after which many uphold such vows as to have no more children or to live off alms (see also Hanssen, this volume).

Because Baul renunciation ideally requires a couple to become initiated together, single Baul women who choose renunciation are faced with the double challenge of leaving society's roles for them as women and of going against Baul expectations of what religious practice means. Baul women I met frequently expressed concern about the importance of being a wife both in Bengali society and in Baul practice. Their stories revealed efforts to guard their reputation through displays of feminine respectability and sexual modesty, concerns also shared by other women renouncers (see, for example Khandelwal 2004; also see Vallely, this volume, for a discussion of Jain ascetics who aim to become detached from their sexualized bodies). But despite the challenges of being single women renouncers in a couple-oriented Baul tradition, many who were alone were ambivalent about the advantages of having a husband.

In this essay I discuss the life stories and perspectives of Vaishnava and Muslim Baul women as they contemplate and explain their choices regarding renunciation.[2] I argue that taking renunciation as single women in a couple-oriented Baul tradition is a way to respond to concerns about their otherwise problematic gendered identity and status within the larger non-Baul society. Pursuing renunciation as a way to deal with societal concerns (as opposed to being exclusively or primarily spiritually motivated) is not unusual, nor is it exclusive to single Baul women. Married and single Bauls very often use renunciation to negotiate their position and identity within society, or to marry a person deemed unsuitable by family or community. Hausner (this volume) also shows that *yoginis* may take renunciation to evade marriage or to leave an unhappy home.

For Baul women who are expected by both mainstream Bengali society and by Bauls to be married, renunciation can help legitimize their single status. They demonstrate that as renouncers they are no longer viewed merely as unmarried women, widows, or wives abandoned by husbands. Thus by aligning themselves with larger renunciant traditions, these women—so precariously situated between ideals for women and ideals for Bauls—gain some prestige and legitimacy implicit in being a *sannyasini*.

Furthermore, as I attempt to show below, these women reveal that they are not renouncing society at all, but rather society's expectations of them. For most of the women I interviewed, renunciation did not precipitate any severing of social ties, but because it did loosen society's claim on them, it provided them with an excuse for evading societal roles and expectations. As single women alone they were then able to pursue their musical and/or spiritual path and justify their mobility in public. For these women renunciation is very much an act of agency, and they demonstrate an awareness of their gendered positions as they creatively interpret their lives and their options.

The material for this essay is based on ethnographic fieldwork conducted between 1998 and 2000 in West Bengal and Bangladesh. The women I discuss in the following sections come from Birbhum (largely around the small university town of Santiniketan), a district in predominantly Hindu West Bengal, India, and the cities of Sylhet and Dhaka in predominantly Muslim Bangladesh.[3] Vaishnava (Hindu) and Muslim Bauls generally have common philosophies and practices, though the images and words (such as Krishna or Allah, the Vedas or the Qur'an) they use to discuss their views may differ. Nonetheless, there is a great deal of cross-fertilization with songs, language, and lively discussions shared among Bauls from different backgrounds. While perhaps unexpected and controversial among those who see Baul sexual practices between a male and female partner as central to the definition of Baul, I focus here largely on single women in order to highlight some concerns many Baul women, coupled or not, expressed when discussing renunciation. All these women did consider themselves Bauls, whether they had a partner or not.

Expecting Renunciation

In the popular imagination of Bengalis and visitors to West Bengal and Bangladesh, Bauls are generally depicted as a sect of musical mendicants clad in ochre-colored clothes who sing about their quest for the Divine within each human and espouse an egalitarian and anti-sectarian view of society. Their numbers tend to come from the lower castes and classes, and Muslims and Hindus both become initiated as Bauls, relinquishing in the process a part of their previous status-conscious identity. They are often described as *pagal* or *khepa* (mad) and their unconventional and mad behavior is understood as being due to their intense longing for the Divine. Those who have read about Bauls also know that Bauls engage in sexual rituals and the ingestion of potent bodily secretions—practices (*sadhana*) deemed important by Bauls, as discussed by Hanssen (this volume), but highly problematic by many Bengalis. Despite such general conceptions, often even well informed, coming up with a consensus about Baul identity is actually an impossible task.[4] Questions of Baul authenticity are sometimes linked with renunciatory status, and many Bengalis in the Santiniketan area where I did much of the research represented in this essay expect that Bauls have taken renunciation.

In particular, I heard many elite Bengalis and Bauls claim that *madhukari*[5] (begging for alms) was a clear sign of a real Baul. What makes *madhukari* so important in the eyes of Bengalis is its association with renouncer traditions. This is partly due to the idealistic stature attained by Bauls in this area, such that many Bengalis prefer to connect Bauls with the more prestigious *sannyasi* traditions. Indeed making this connection places Bauls in a more favorable light than admitting to their unorthodox sexual practices. Furthermore, many Bengali Hindus maintain that giving alms to those who have fully devoted themselves to a religious path is a duty of householders and will bring merit to the donors. Many renunciant traditions follow strict guidelines for behavior and require proof of lineage to demonstrate the authenticity of one's status as a legitimate renouncer. For some Bengalis, this emphasis on renunciation and *madhukari* is an attempt to separate so-called real Bauls from artists (*silpi* or *gayak*) who rely on contributions and payment at programs. The underlying assumption here is that only those who have taken renunciation are entitled to live from the alms of *madhukari*, and those Bauls who have renounced are seen as having taken the necessary departure from worldly concerns and greed. In contrast, however, those Bauls who do not do *madhukari* are considered to be Bauls only for the gain of wealth and fame and are frequently believed not to be real Bauls.

Most Bauls I met, however, never really seemed to emphasize renunciation, and my observations led me to believe that although some of them had taken renunciation, their lives did not seem extraordinarily different from those who had not. In fact, there appeared to me to be an implicit ambiguity both in one of the Bengali terms most used for Baul renunciation and the external markers of a renouncer. Baul renunciation is typically called *bhek* (and having Vaishnava associations) or *khilaphat* (the term used mostly among Sufis, though Muslim Bauls use the term *bhek* as well). Other terms also used are *sannyas*, which has its association with the larger sadhu community (sadhu *samaj*), and *tyag*, which means giving up or relinquishing, and the Baul renouncer may variously be referred to as *bhek-dhari, bairagi, sannyasi*, or *tyagi*. The term "*bhek*," which was most often used among my informants, also refers to the attire of an ascetic, particularly a Vaishnava ascetic, and the putting on of that attire. In the eyes of the uninitiated public, these understandings are often conflated so that ochre-colored garments indicate a renouncer. Bauls frequently told me that they wear ochre-colored clothes, regardless of whether or not they are renouncers, because it is expected of them in society (particularly when performing). Even Muslim Bauls and Vaishnava Bauls who tend to wear white instead of ochre are likely to wear ochre when performing in public because much of society (even in predominantly Muslim Bangladesh) connects that color with renunciation and believes that renouncers are entitled to beg for alms, as Bauls are typically perceived as doing when they perform.

Among Bauls, there are conflicting and ambiguous views on the importance of taking renunciation. Part of this can be related to the socioreligious contexts in which particular Bauls find themselves, even though sectarian differences are downplayed in Baul discourse. Thus among Muslim Bauls, renunciation is less practiced than among Hindus, and Muslims in general tend to applaud householder life and responsibilities over renunciation. Vaishnava Bauls in the Santiniketan-Bolpur area are more likely to talk of the value of *bhek* (though many may never take it) in part because the Hinduization of Bauls in the area means that many (especially, it seems, the Bengali elite) use renunciation as a measure of authenticity.

Although the socioreligious environment may color to some extent the way Bauls view renunciation, there is no consistent perspective on the matter among Hindu and Muslim Bauls. This is largely attributed to the diversity of the Baul path, a characteristic shared by other South Asian religious traditions, despite scholarly (and lay) efforts to determine a uniform perspective. Although certain gurus have their ideas about the best way to progress, there is no uniform Baul path, and a majority of Bauls are householders regardless of the level of their practices and expertise or their renunciant status. Thus renunciation, though it may be viewed as an ideal way to relinquish societal obligations and devote oneself more fully to *sadhana*, is not particularly seen by Bauls as a culmination of one's spiritual endeavors.[6]

I would suggest then that the motivations for taking renunciation among Bauls with whom I spoke differ from a more conventional view held by scholars and householders that renunciation is a step intended to cause the realization of illusion (*maya*) through severing ties permanently with one's family, community, and responsibilities. In contrast, Bauls who take *bhek* can and often do continue with their relationships in much the same way as before. Renunciation for Bauls has a distinct flavor that sets itself apart from mainstream paradigms of renouncers whose lives are characterized by celibacy, asceticism, and uninvolvement in worldly concerns. Most Bauls would agree that renunciation is ideally taken as a couple with the purpose of pursuing their *sadhana*. In fact, most would severely criticize the practice of taking renunciation as an individual and would consider any individual alone not as a genuine Baul because of the centrality of *sadhana*. Followers of the famous nineteenth century Baul Lalan Shah (*Lalan panthis*) in Kusthiya, Bangladesh, are adamant that an individual cannot take *bhek* alone (Caudhuri 1997).

Despite the strong stance taken by many Bauls on this issue, there are individuals who take renunciation alone and who do consider themselves Bauls. Most of the women I focus on in this essay fall into this unusual category and have taken renunciation alone. In fact, as their stories show, renunciation has been a way for them to mitigate their otherwise problematic status as Bengali women alone.

Stepping Out Alone

Taking renunciation in many South Asian traditions involves the often difficult act of removing oneself from normative householder roles. A woman who takes this step typically goes against society's expectations of her by leaving traditional gendered roles, particularly those of wife and mother. Women who take renunciation without a husband in an otherwise couple-oriented Baul tradition, however, are further challenged by the fact that they do not follow Baul expectations of ritual practice. The single Baul women renouncers with whom I spoke were very aware of these tensions, of the problem that the cloak of renunciation under which many of them were seeking protection did not provide them with the solace of spiritual development that comes from partnered *sadhana*. Some women bemoaned the lack of a partner and husband, arguing that they would have preferred to have been a couple. Even those women who blatantly rejected partnered *sadhana* in favor of an independent life without a husband were aware of the tensions between their own circumstances and societal and Baul expectations to have a partner. This ambivalence about taking renunciation, which is shared by many Baul women and is reflected in their statements, concerns, and stories, is grounded in their gendered positionality: their positions both as Bengali women and as Baul women set them up as needing a husband. Even women who consider taking renunciation with their husbands may fear the repercussions of severing ties with their social roles as Bengali mothers and wives, and this is largely due to the potential of ending up alone, that is without a husband and ritual partner. As one Baul woman explained: "For an 'Indian' woman, the husband is everything." By extension, I suggest, so too is her position as a wife. Whether Baul women who take renunciation without a husband do so defiantly or ambivalently, the husband and her status as a wife factor in as major concerns. In the sections that follow, I argue that women's decisions to take Baul renunciation are not so much a reflection of expectations and options available within the Baul community as they are considerations about these women's situations vis-à-vis the non-Baul society in which they live.

Renunciation to Legitimize the Lack of a Husband

Kangalini Sufia and Madhabi are two women whose life trajectory bear some important similarities and reveal the ambivalence some Baul women have about the importance of having a husband around. For both, renunciation became a way of responding to being young women and mothers with husbands who were no longer present. And indeed they took renunciation after their husbands left, in contrast to more expected Baul patterns. But while Kangalini's life story does suggest some dramatic breaks from her past identity, life, and family, Madhabi remains closely connected with her parents and daughter.

Kangalini Sufia's story about taking renunciation and becoming a Baul is in many ways typical of single Baul women. According to the version she told me, her marriage was arranged when she was around age seven, and she had a daughter by that marriage. Her husband used to beat her and treat her poorly so, after much difficulty, she returned to her parents' home. Her father and brother were killed during the Liberation War of 1971 with East Pakistan because they were Hindus. Her mother brought Kangalini and her infant to a Vaishnava renouncer (*bairagi*) who promised to take her in and give her some instructions. Under this guru, Kangalini took *bhek*; she was probably 15 at this time. Later, she took a guru who instructed her in Tantric practices as well as songs. It is from him that she says she began learning Baul songs, and after leaving from his care she took the path of a Baul.

At some point she became a Muslim and married a Baul.[7] "When I was a Hindu I had a lot of troubles. Later on I became a Muslim. Then I learned the Islamic articles of faith (*kalima*). Then I became happier. Then I could stand on my own two feet. Then I could go out and make some money. Then I was able to be a little happier." Today she is one of the most well-known Baul singers and songwriters in Bangladesh, but for Kangalini becoming a Baul was very much a strategy for coping with being a young woman alone, that is, without a husband or family to look after her. As a Hindu, remarriage was not a viable option once her first marriage failed, and since her natal family had already suffered enormously during the war, they could not provide her and her infant with shelter. Kangalini claimed that it was precisely because she was a woman that she became a Baul. "If I was a man, I would stay in their world. Whatever my family occupation (*jati byabasa*), I would just have done that." Taking *bhek*, however, gave her status as a single woman some legitimacy and entitled her to live off of alms. Kangalini, however, was quick to explain that this was not enough: being a Hindu in Bangladesh at that time was a source of difficulty, and hence she eventually became a Muslim. Unquestionably Kangalini has great talent in singing and creating songs and thus having a career in performing Baul songs proved possible. Nonetheless, taking renunciation, becoming a Baul, and later a Muslim can be seen as attempts at easing some of the challenges she faced and, significantly, ones about which she was very conscious.

Although it is more expected in renunciant traditions to leave one's spouse, family, and community in order to pursue a life of religiosity, Kangalini became a renouncer as a way to respond to her life as a single woman. Her ambivalence about this past can be viewed more clearly when comparing her to another woman who actively and defiantly severed ties with her marital life. Like Kangalini, Rajani took *bhek* after leaving her husband. Rajani is an older woman living in an ashram in a town in Birbhum, West Bengal, who casually refers to herself as a Baul ("Baul, Vaishnava: they're the same thing"). But their memories of why they left marriage are articulated quite differently: whereas Kangalini left because her husband was abusive, Rajani deliberately left her husband in order to pursue her spiritual path. It is likely because of these different reasons as well as the divergent directions their lives have taken—one performing on big stages both in Bangladesh and abroad, the other living in an ashram and singing for alms—that the stories they tell of their previous marriage demonstrate radically different ideas about appropriate wifely behavior and responsibility toward one's husband. Rajani claimed that it was an independent life that she wanted all along and that she was frustrated that her husband prevented her from doing things she wanted to do, namely singing and pursuing her religious path. Since she had no children from her marriage, she was able to break ties with her husband, encourage her in-laws to find another wife for him, and then leave that life behind. In fact, she described leaving her husband with some pride, demonstrating her defiance against the traditional wifely role given to her by exclaiming, with a gleam in her eye, that it was *she* who had left *him*. In contrast, Kangalini is conflicted about what she should have done in her previous marriage. "Maybe I was at fault," she suggested. "Maybe I did something wrong. Now my husband is very reddish—like you. His skin was light skin. I was black. Maybe that was my fault. I was black."[8] In most versions of her life story, Kangalini claimed that her husband died and that it was as a widow that she began her life as a Baul. This was the version she first told me as well, but during a much later interview, she changed that statement and said that he is still alive and that she had left him because he was so abusive. Furthermore, there was no defiance in her voice when she talked about how she left. She said that she tried to stay, tried to do everything right as a wife, but that he continued to abuse her. In one interview, she claimed that, "[If I knew then what I know now] I would've grabbed my husband's feet and stayed. Even getting beaten, I wouldn't have left his house. But then I didn't understand this! I didn't know." Kangalini, then, did not readily articulate defiance against the dependence of women on men but at times stated that it is women's duty to fulfill such roles:

> In the Qur'an Sharif . . . they know that by serving the husband, [one] can get close to Allah and go to heaven. You have to take care of the

> husband: wash his feet; dress his hair; wipe his feet. Only loving him privately is not enough; you have to show how you love him. You have to give him all kinds of things. You have to give him whatever a woman has. Even if a husband beats her, you can't counter him. Wherever a husband puts his hand on a woman, that place is heaven.[9] All this we [women in general] didn't understand before. If we understood this, I think that a lot of women would stick with their husbands no matter how they were abused.

Unlike Rajani and many other women renouncers who have sought out solitude and refuge in ascetic devotion, rejecting in the process their societal role as wife, Kangalini is a woman who performs onstage and depends on her image for her living. She therefore attempts to construct her own identity as a good wife in order to maintain an image of respectability in the eyes of potential patrons and audiences. Her life as she described it is marked by struggles concerning her precarious and problematic positions—as a young mother, a Hindu, and a single woman—and in her story she articulated numerous attempts at negotiating her identity and position in society. These attempts at negotiation not only occurred in the past, however; they continually occur as she tells and retells her story. For instance, she tells many people that she is a widow instead of revealing that her husband is alive and had abused and rejected her. Many Bangladeshis, even Muslims for whom divorce and remarriage are religiously sanctioned for women, would agree with her statements above that a wife should remain with her husband no matter what.

I believe that Kangalini's purported support of mainstream roles for women is in part due to her experience that taking the path of a Baul as a woman is extremely difficult. This is indeed evident in the sentiment expressed in the very beginning of our first interview: "My Baul life is a life of pain" as well as many other descriptions of her life as a Baul. Although she has been very successful as a Baul singer, her statements suggest that she believes it may have been easier to have remained a traditional wife, even if married to a man who abused her.

Another Baul women, Madhabi, I met through one of my main informants, Jaya, outside of Santiniketan where I had my base for much of my research in West Bengal. Jaya had been trying to explain to me the challenges she faced as a Baul woman, and she introduced me to Madhabi in order for me to understand how difficult life can be for a young Baul woman.

Madhabi arrived sitting on the back of a man's bicycle to meet me at Jaya's home. She was a small woman in her late twenties and was wearing a bright ochre-colored sari with a string of *tulsi* beads around her neck and a longer beaded necklace. The man with whom she arrived, I later learned, was her fictive older brother—her *dada*—and he too was clothed in ochre: a long lightly colored ochre *panjabi* (tunic) shirt and a white *lungi* (a long cloth tied at the waist). As soon as we sat down on Jaya's verandah, Madhabi launched into her story about her "very difficult" life. I had to scramble to pull out my tape recorder, but as soon as she saw it she appeared eager to record her story into the machine. Her story, as she told it, began right at what she seemed to think was the crux of the problem:[10]

> My husband told me that my family would have to sign the ashram over to him, but my father answered, "If I sign it over to you, then my daughter will have no future. My daughter will have a future, so that signing over will not happen. You are a *gharjamai* [son-in-law who lives with the in-laws], you're fine; your needs will be met by staying here with us." But my husband wanted the ashram to be put in his name and did not want to be a *gharjamai*, so he said he would leave. After he left, my parents did not know what to do with me. I stayed with my uncle and aunt, but they died soon after. After that I had to arrange my daughter's marriage. Once that was done, I told my father that I would go stay with *dada*. As I said from the very beginning, if my husband leaves, I won't marry again. If there wasn't happiness in this marriage, why would happiness come from the next time?

She told me about how her husband used to beat her, drink alcohol, and then finally left her and their daughter because her family refused to sign over their ashram to him. Although he was a *gharjamai*, having his basic needs met by living in the home of his in-laws, the more expected pattern in South Asia is for the wife to live in the house of her husband and in-laws. I do not know why Madhabi was not living with her husband's family (perhaps he was an orphan), but it is likely that due to this irregularity, he viewed being a *gharjamai* as a weakness and therefore wanted to have the ashram property put into his name. As her parents later told me, they had not signed the ashram over to her husband because they were convinced that he would sell it for cash, leaving them all without a home. They too claimed that Madhabi's husband had turned out to be no good. Though Madhabi and her husband had both taken initiation (*diksha*) together when they got married, receiving from their guru a mantra for male and female partners (*yugal mantra*), she took *bhek* when he left. When I met her, she was staying with her *dada* and his wife and children in Bolpur where she was hoping to earn enough money by singing and begging for alms to keep herself going.

It was monsoon season and the rains forced us inside. During lulls in the storm, she would prompt me to turn on my tape recorder again so that she could tell me more. Madhabi spoke with such urgency that I felt all the questions I had prepared were painfully inadequate. Even though her husband had left three years ago, her voice had the urgency of new pain as she called him a scoundrel (*badmas*) and described herself as having done everything right as a wife.

I began meeting her at her *dada*'s and we continued these conversations. During these meetings, she would wear an old white sari and no ornaments save a garland of beads made from basil wood (*tulsi mala*). Many days she was running behind in her morning duties, so I would sit and wait as she returned from her bath, and I watched as she offered incense to the deities depicted in posters in a corner and to whomever was sitting in the room. As she encircled me with incense, she offered the familiar Baul explanation: "We worship humans. It is important to love humans first." One day she described to me her *bhek* initiation, explaining taking off her marriage bangles, her guru giving her a begging bag, and then going to some homes in the village to beg for alms, and finally also receiving alms from her own parents: "Oh how they cried! They cried so much!" she said. My training told me that the act of begging from one's parents was surely a ritual leading to an actual severing of parental-child ties, demonstrating that she had renounced family life and responsibilities. Assuming that her parents' grief indicated such a dramatic transformation, I offered the next question: "So after you took *bhek*, your relationship with your mother and father changed?" But my question did not lead to the answer I was expecting: "I now have four mothers and fathers. I have one set more than before. When I was given in marriage, I had another set of parents [for a while]. They [too] were my gurus. First guru . . . second guru. My mother and father were the first parents, then there was second [in-laws], [then] third [*bhek* gurus]." On the one hand, Madhabi was explaining that one's mother and father are also gurus, as parents are for all children. But the reason she dismissed my suggestion that her relationship with her parents had changed became clearer as time went on: except for their increased concern about Madhabi's unhappiness, their relationship had *not* really changed, at least not in any way that I could perceive. Describing a similar scene in which neighbors grieved at the renunciation ceremony of a young Baul woman, Hanssen (this volume) points out that the community associates this act with a severing of ties, even if they know that no such actual break will occur. Madhabi's parents grieved, but they knew their daughter was not lost to them.

Madhabi frequently punctuated our conversations by saying she would take me to see the ashram, which she and her parents had refused to sign over to her husband. One day we made the three-hour bike ride to the ashram, accompanied by her *dada*. We cycled for some time along the road that follows the railroad track, and then we veered off the main road onto a winding dirt road. On both sides were fields, and villages appeared in clusters of trees scattered along the horizon. The small road we traveled on did not appear to lead to any of these villages but persisted in veering away from them. At one point, however, we stopped, and Madhabi pointed to a cluster of trees in the distance. There, she said, was where her family's ashram was; that was where we were going. The distance would not have been so far except that the road we were on did not lead us in a straight trajectory to our destination. Clouds were beginning to accumulate, giving us some relief from the sun, but the rain was imminent.

We reached the trees of Madhabi's natal village just after the rain began. Madhabi greeted people running across our path as we hurried to her parents' home, umbrellas in one hand, and leading our bikes with the other. Like many villages in Bengal, this one was lush with vegetation: palm, bamboo, banyan, and small shrubbery. We took a smaller path to an isolated area of the village and arrived at the ashram. There was a small shaded pond on one side, fruit trees near the house, and four graves[11] made of pressed mud. The mud house had two stories, but it was in need of repair, with portions of walls beginning to crumble. As we approached the house, we saw several people: Madhabi's parents, who live in the house, and her daughter and son-in-law who live in another village and came for the day to visit. They were rushing around, pulling things out of the rain, and unrolling tarps that were attached to the roof of the verandah. When he saw us, Madhabi's father waved us into their home as we abandoned our bicycles and sandals and rushed inside, already soaking wet. Madhabi's daughter handed us dry towels and clothes as her parents attempted to secure the tarps around the verandah against the relentless rain. We sat at first on the verandah, but soon the wind forced the rain through large rips in the tarps and we all crowded into a tiny dark room shared by small images of Radha and Krishna.

I met her parents only once, but I got a sense of the extent to which their lives were closely entwined with their daughter and granddaughter. These were not people who had renounced family obligations but rather tried to maintain family life alongside their own *sadhana*. Madhabi's parents told me they had taken *bhek* when they learned that they were going to have a child. They did this because they felt they had to demonstrate publicly their commitment to the Baul path, and *bhek* indicated a vow that they would have no more than this one child, Madhabi. When I met them, they were living alone in what they called an ashram and were subsisting on *madhukari* in their village. They claimed their neighbors knew nothing of their sexual rituals, preferring instead to see them as more orthodox Vaishnava renouncers. Although their identity as renouncers improved their status in the village and allowed them to live off alms, Madhabi's parents claimed that taking *bhek* was particularly important for their reputation as Bauls in the Baul community. But *bhek* by no means implied a separation from their duty as parents. Madhabi's parents' commitment to their daughter extended to wanting Madhabi to be a Baul, a path they believed should have given her peace and spiritual growth, which is why they had originally arranged her marriage with a Baul. Madhabi's marriage failed some years after the birth of their one child, a daughter, so that raising the child, providing her with some education, and arranging for her dowry and marriage fell on Madhabi and her parents. Although when I met them they expressed some satisfaction in knowing that Madhabi was living with her *dada* and his family and was in a sense protected, they still preferred to find another husband for her because without a partner, they insisted, she could do no *sadhana* and would therefore never really gain the peace that they had hoped for her.

Madhabi and her parents took *bhek* for very different reasons. Although Bauls regularly argue that the same goals can be accomplished without formally taking *bhek*, many Bauls view renunciation as an opportunity to liberate themselves from societal obligations, giving them the space to focus on *sadhana*. It is in line with these objectives that her parents took *bhek*. In contrast, Madhabi took *bhek* alone, and her motivation for doing so was her husband's leaving her. When I asked her how she came to the decision of taking *bhek*, she answered: "What will happen to me? For that reason." Several times throughout our conversations, she stated that as a woman, she was very limited in what she could do:

> There's a need for a man. If you go out alone, things happen . . . It happens, right? So [people say to me], "You don't have to do all that Baul stuff. Stay at home. Where will you go all by yourself? [You're] a woman. Traveling all over, long distances. Alone, there's danger." . . . Anything can happen. . . . If you don't have a man, [then as a] woman [you] can be maligned. Won't it happen?

As unconventional as their original motivations may have been, both Madhabi and Kangalini show that something is gained by taking renunciation. As renouncers they are able to justify their status without a husband. They are able to travel about alone, even though, as they both suggest, they do face difficulties without the protection of an accompanying male. But, as explained by Madhabi, they can draw on their status as renouncers in order to defend their situation alone, and oftentimes this initiates the protection of householders who see it as their duty to support those who have devoted their lives to religious pursuits. Finally, renunciation gives them the public sanction to live off alms, giving them at least a little from which to live.

Kangalini has made a break from her past life and identity, but renunciation is not particularly the reason why. According to Kangalini, there is no one left in her family except for her daughter; her parents and brother all died long ago. Although she has made a break from her previous caste affiliation, she articulates this as being due to her conversion to Islam, not renunciation. Madhabi and her family further demonstrate that Baul renunciation does not indicate a dramatic transformation in their status as householders or as people involved with worldly concerns. Since Baul renunciation does not necessitate breaking off ties with family and community, as it does in most other South Asian traditions, what then is changed or renounced when a Baul woman takes renunciation? In the following section I address this question by considering the ambivalence about Baul renunciation expressed by Madhabi and Jaya. I suggest that their apprehensions about becoming renouncers help us to understand what it means for a woman to be a Baul renouncer. This will in turn shed more light on the larger puzzling question of why to take renunciation when so little appears to be actually renounced in the process.

The Value of Living in Society and Being a Wife

The more time I spent with Baul women, the more evident it became that there is a significant tension between having—or wanting to have—a husband around and not. Baul renunciation poses particular challenges for women because of both the Baul and the societal importance of having a husband. Bauls are not only ideally supposed to take renunciation together as a couple, they are also expected to remain together in order to receive true benefit from their *sadhana* (promiscuity is said to decrease the effectiveness and worth of *sadhana*). Although ideal Baul expectations are that couples remain together, their status as Bauls and particularly as renouncers removes them from many of the local societal expectations and pressures to remain married. As a result, it is not uncommon for Baul men to leave their wives (see also Hanssen, this volume), and many women I met believed that Baul renunciation even taken as a couple potentially jeopardizes their roles as wives. It is largely for this reason that Madhabi did not want her daughter to become a Baul and why Jaya, who introduced me to Madhabi, stated that she does not want to take *bhek*.

Madhabi and Jaya live in Birbhum near Santiniketan, where Bauls have a reputation of having impermanent marriages. When Jaya first suggested I talk to Madhabi, she described her as a renouncer who was having a very difficult time because her husband had abandoned her. Jaya, a married Baul woman in her mid-twenties living with her husband and daughter, lives very much in society. She is a young performer of Baul songs with a growing commitment to Baul philosophy and *sadhana*, and she says she never wants to take renunciation. At the time she suggested I meet Madhabi, she was trying to explain to me the challenges she faced as a Baul living in her village when much of what she believed as a Baul went against the expectations and views of her neighbors. As Madhabi's own story unfolded, discussed above, she too articulated her husband as the source of her problems.

On the day that Madhabi took me to her parents' ashram where I also met her daughter and son-in-law, we sat on a hill overlooking a pond. Her daughter was around 15 with long thick hair loosely draping down her shoulders and big doe-like eyes. Her responses to my questions were usually monosyllabic and uttered quietly, but Madhabi, who had wanted me to interview her daughter, would frequently instruct her in what to say or jump in with her own comments and clarifications. I learned very little about her daughter that day, but I learned quite a lot more about Madhabi. When I asked the daughter if she had learned Baul songs, her quiet "no" was followed by Madhabi's interjection that being a Baul is not such a good thing, suggesting that she had not wanted her daughter to be a Baul. Up to that point, Madhabi had in our interviews kept two narrative threads somewhat separate: on the one hand she was miserable because she had a husband who was abusive, irresponsible, and selfish, and on the other hand she found peace singing Baul songs. If there was any connection between the two, it was that being a Baul and singing Baul songs had been a way for Madhabi to cope with her circumstances: singing provided a small income and offered her some peace of mind. But usually she articulated a contrast between the peace (*shanti*) and happiness (*ananda*) she gained from singing Baul songs and being able to move about and mix with people and the lack of peace in her marriage: "This marriage has been really unpeaceful (*ashanti*)."

When I asked her daughter if she learned Baul songs, Madhabi revealed a causal relationship between these two aspects of her life: her own misery was in part due to being a Baul. That is, rather than emphasizing the peace that can come from the Baul path, she revealed that being a Baul can actually bring one misery because a Baul woman's husband may leave. Because of the precariousness of Baul marriages, Madhabi wanted to make sure her daughter would not become a Baul, and she did this in at least three ways: First, she did not teach her daughter Baul songs, saying instead she should go to school. Although she was only in school up to grade five, at which point her marriage was arranged, Madhabi's daughter was given the message that schooling and being a wife were more important than singing. This contrasts with Madhabi's own childhood during which she enjoyed and participated in singing with her parents at home, in addition to receiving some schooling (also up to grade five).

Second, she made sure the daughter married a good man, and this was accomplished in part by providing a dowry and wedding ceremony. Madhabi's many articulated worries about getting her daughter married properly demonstrate her determination to make sure her daughter did not have the kind of life she had to lead. Thus Madhabi explained that her family sold part of the ashram land to pay for the dowry: "I gave 24,000 rupees . . . I counted it and gave it to my daughter. On my daughter's ears, throat, wrists [I placed ornaments]. . . . [And] I gave a wedding." The wedding, she described, was with a priest and the appropriate amount of fanfare, both of these being needed to demonstrate her family's commitment to the daughter and provide some degree of insurance that the bride will be respected in her in-laws' home. Furthermore, bringing me, a foreigner, with all the prestige and power implicated in my mere presence in her country, to her parents' ashram and arranging for me to meet her daughter and son-in-law (and then asking me to interview her daughter) further enhanced her daughter's position. Although I had often felt like I had been "shown off" by my interviewees to others, I realized later that my presence was at least partly a substitute for her family's lack of other resources for providing gifts for the daughter. During the day I was there, Madhabi often complained that she had nothing to give her daughter and that her daughter's father never gives anything—money, food, nor gifts. Such gifts are important in demonstrating the natal family's continued support of a daughter, further ensuring her well-being in her in-laws' home. Madhabi pulled me aside toward the end of our visit to ask if I could give her daughter some money for a sari. I had not gotten the hint embedded in her complaints about her own lack of financial resources.

The third way Madhabi attempted to ensure that her daughter will not become a Baul is by not remarrying. Although she could draw on her freedom as a Baul to remarry, Madhabi instead emphasized her position in Hindu Bengali society and stated that remarriage would look bad in the eyes of that society, negatively affecting her daughter and in-laws. So she claimed: "I won't marry again . . . I have a daughter, a son-in-law. It's a 'prestige' thing. In our society, two marriages are a matter of 'prestige.' Scoundrels, they do that." When I mentioned to Madhabi in another interview that Bauls frequently do remarry, she agreed and told me that those that do are scoundrels. Madhabi could not deny being in the Baul path herself, and the marginality implicit in being in this path was further heightened because her husband had left her, leaving her alone and without the protection of a husband. But even despite these factors, she still tried to demonstrate the many ways in which she had done right as a Bengali woman. And besides providing for her daughter's marriage, this includes not remarrying. But it also includes taking *bhek*.

Madhabi took *bhek* in part because it served to legitimize her position alone, a decision motivated by her options in the larger Bengali society, not so much among Bauls. Baul remarriages are fairly common, and Madhabi would have had the support of her parents to have married again. Madhabi's parents' concern for their daughter's solitude was articulated to me primarily in terms of her lack of a partner for *sadhana*, a practice that they believed would give her peace of mind as well as essential spiritual knowledge. Although her parents did not express it in our interview, I feel certain that they were also concerned about her not having the protection of a husband. Madhabi frequently talked about how difficult it was to be out in public without a husband, and living with her *dada* and his wife only eased that problem slightly. Furthermore, because her parents had some ashram property, they would prefer that Madhabi had a supportive and trustworthy spouse so that they could leave the property with both of them and not worry that her previous or a future husband might abandon Madhabi again and sell the property. Madhabi said she would get the property herself if alone, but she doubted that there would be anything left by the time her parents died. Since there were no young males contributing to the household, any difficulties her parents may face, such as illness, will most likely have to be met by selling off remaining portions of the land. If Madhabi had a husband who was willing to help maintain the household, the two of them would probably be able to keep the ashram after her parents' deaths.

Thus there are many reasons why Madhabi could, as a Baul woman, remarry. I would suggest then that Madhabi's decision to become a single renouncer reflects more her awareness of her position in Bengali society than it does her affinity to the Baul community. As a young single Hindu woman abandoned by her husband (who has since married another woman), her identity is problematic. Literature on Hindu women frequently stresses the importance of a husband to secure the status of a woman, and a woman who has lost her husband, such as a widow, has also lost the valuable role of wife (and sometimes mother).[12] Madhabi's husband failed her, not because he died but because he left. A husband's death or abandonment is equally problematic, and Madhabi revealed a common perception, often attributed to Hindu widows, that her misery was likely her fault: "In a past life I did something," she said. After her husband left, her social identity as a wife was shattered, and she turned to her responsibilities as a mother to look after her daughter's well-being and marriage.

Madhabi took *bhek* because she considered it the best option for both her and her daughter, especially in light of her own experience of the precariousness of marriage. Taking *bhek* gave her single status some legitimacy in Hindu society, conveying the message to her son-in-law's family that she believes marriages to be final. What Madhabi will do when her daughter successfully bears a child—especially if it is a male child—remains to be seen. I would speculate that once a child is born, Madhabi will feel her daughter's security is more assured, thereby freeing up Madhabi to make a different choice—possibly remarrying.

One message that Jaya was trying to get across to me by introducing Madhabi was that Baul women are vulnerable in ways that Baul men are not. Jaya knows that her husband could also leave her, and she is torn about the question of taking *bhek*. On the one hand she says that she would like to take renunciation because she really wants to move to another level and learn more from her guru. She would also like to be able to sit with her guru and other *sannyasis*, sharing meals and conversation with them. But Jaya also is very concerned about her and her daughter's future. Because she has seen so many Baul men leave their wives and family, she knows that her husband could easily do the same. For that reason, Jaya has decided, for now, not to take renunciation and to remain instead symbolically within society. By doing so, she hopes that her neighbors (who are not Bauls) will give her the protection she might need in the event that her husband does consider leaving her. Jaya therefore also attempts to maintain a good reputation in her community, and this effort is largely a response to her fears of abandonment. So when I asked her if she wants to take renunciation, she explained:

> I live here. I won't leave from here [community]. So I have to follow the community. I said one time that if [my husband] ever leaves me, then the community will look after me. My community will keep me safe/fine. But if I leave society now, if I go against society [if I renounce], then later it will be very difficult for me. Community is another level of existence. Whether we agree with it or not, I have to live in that society.

Jaya expresses a tension between being in society (*samaj*) even while on the Baul path and stepping out of society's realm of protection. Her statements suggest that even as a Baul woman, she can maintain an image of living in society, cultivating the ties of attachment (*maya*[13]) with her non-Baul neighbors. She and Madhabi both say that so long as they behave correctly, they will remain in good light and receive protection. In different ways, they both try to assure society's protection of them. Renunciation for Jaya and other married women would involve removing themselves from the protection of society. At the same time, renunciation can offer single women another type of protection: the protection that comes from being a *sannyasi*. As a *sannyasi*, she enters a receiver-donor relationship with Hindu householders. For a woman abandoned by her husband, like Madhabi, this option provides a semblance of normalcy to her obviously aberrant position as a young single mother. In rural Bengal, her identity as a single mother would be uncertain; as a Baul renouncer her identity is as a *sannyasi*, and she can evoke local understandings of the obligations and responsibilities householders have to renouncers to her benefit.

"Back Then I Had No Sense": Severing Ties as a *Pagal*

As demonstrated above, renunciation on the Baul path, intended to be taken by a man and a woman together, is viewed by some women as threatening their social role as wife and sometimes mother. There are other Baul women, however, who are eager to sever ties with the mundane world and would happily abandon conjugal life.[14] In their efforts to live a religious life, they show that for some women, marriage and religiosity are not compatible. Rajani, whom we met earlier in this essay, was able to turn to the *sannyasi* tradition as a way out of her socially prescribed role of wife, making sure ties were severed with her husband and in-laws by insisting they find another wife to replace her. As a result of becoming a renouncer, she was able to focus all her time and energy on her *sadhana*, unencumbered by family obligations. Nur Jahan drew, ultimately with less success, on other options for evading the mundane world of marriage.

Nur Jahan is a Muslim Baul woman I met in Sylhet, in northeastern Bangladesh, in 1999. Although she did not take formal renunciation, her story is relevant here because she did leave society in a similar way, and the challenges she faced while caught between worldly concerns and a life of devotion also bear important resemblances with other women discussed in this essay. As a Muslim woman, the institution of renunciation, more readily associated with Hindu traditions, is not really an option for her. Some Muslim Bauls do take renunciation (usually called *khilaphat* or sometimes *bhek*), particularly among followers of Lalan Shah in western Bangladesh where I conducted some of my research. But Nur Jahan's association with those Bauls, whether Muslim or Vaishnava, who take renunciation appears negligible, and in this and other ways Nur Jahan differs from other Bauls discussed in this essay. Although Rajani met resistance from her in-laws and husband when she left conjugal life, there was an established institution of renunciation she could join that as a Vaishnava Baul she would have been familiar with. Nur Jahan, however, probably would not have considered taking formal renunciation largely because her day-to-day experiences would not have presented her with the option. It is much more likely that she had contact with Sufis who, when they desire to devote themselves fully to religious practices, rarely formally abandon conjugal life even though they may still attempt to avoid marriage. Given the Muslim emphasis on marriage and family life, it is not surprising that those who opt for a spiritual path rather than conjugal life would resort to a variety of different tactics to avoid marriage. When a Muslim saint, spiritual teacher, or mystic is unmarried long past typical marriage age, I have often heard it stated that he has merely not found the right partner yet. Munibor, the Baul singer who introduced me to Nur Jahan, explained that Shah Jalal, the Muslim saint for whom the main shrine (*mazar*) of Sylhet town is dedicated, had intended to marry but died unmarried at an old age because he had not found the right woman yet. I heard the same explanation given for Munibor who was unmarried and in his mid-forties when I met him.

Although I do not have comparable examples of single women in Bangladesh being excused from marriage because "they have not found the right partner yet," Nur Jahan provides an example of someone who tried to remain single. Her attempts, in this case largely unsuccessful, are not without precedence in the Sufi world. Indeed, one of the major early figures in Sufism, Rabi'a of Basra (d. 801), a female saint, refused to get married. Margaret Smith explains that "like her Christian sisters in the life of sanctity, Rabi'a espoused a heavenly Bridegroom and turned her back on earthly marriage even with one of her own intimates and companions on the Way" (1928:13). She is said to have rejected several marriage proposals, claiming in response to one of them that "[r]enunciation of this world means peace, while desire for it brings sorrow . . . God can give me all you offer and even double it. It does not please me to be distracted from Him for a single moment" (Munawi, cited in Smith 1928:11).

Nur Jahan's desire to remain single did not have the effect that it did for Rabi'a, whose historical renderings describe her as spiritually advanced, very influential, and worthy of great respect. It is likely that life circumstances and social pressure did not give Nur Jahan the option of merely refusing marriage in favor of a life devoted to Allah. Instead of taking renunciation, Nur Jahan draws on the Islamic notion of *majzub* or the Bengali *pagal* (mad). As her stories below demonstrate, she clearly sees herself as wanting to abandon her societal roles and responsibilities, particularly those related to being a woman, in order to pursue spiritual goals, and these desires are similar to many of the stated desires of people who do take renunciation.

Nur Jahan's life story, as she told it one afternoon, is full of tensions between different ideals and expectations. On the one hand, she claims to want only to sing her devotion to Allah; on the other, she feels pressure to live in the mundane world, to act in ways appropriate to a Muslim woman.

She refuses to get married; and then she does get married, reportedly under duress. She wants to be close to Allah, feeling no gender distinction in her role as devotee; and she wants to be a good woman, recognizing that this pulls her away from her chosen expression of devotion. Nur Jahan, in the story she told me of her life, never seems to find a balance between the different pulls she experiences, but rather goes from one extreme to another, trying in turn to fulfill her own longing and the expectations she finds others have of her.

It is telling that most of her story was focused on her youth, the years between age nine and mid-20s, the period of time when she was most focused on her devotion to Allah. Now in her 60s, she told me little of the recent 30 or 35 years. Yet, although she focused mostly on those early years, she claimed not to remember them, or at least not sections of them. In a way, it seems as though her claims of not remembering somehow distance her from assuming responsibility for those actions. As she said, she was *pagal* at that time.[15]

After stating her place of birth (near Dhaka) and father's name, she begins:

> People say when I was nine or ten years old, I was of a different kind. I did not learn my lessons. I used to live an abnormal life. For instance, I went down into the water even in the winter. While I was staying in that cold water, my parents sat beside the pond for the whole night. Now, how much difficulties they would endure for me! Then they decided to keep me in chains. Many years passed like this.

Nur Jahan explained that her condition became a matter of honor for her parents especially when it was time for her two younger sisters to get married. Although she was supposed to get married before her younger sisters, she had no interest in "living in this world." She added, "The village council talked to my father seven times: 'Our village's reputation is ruined by your daughter. She visits shrines improperly,' they said. . . . Many religious healers came for my treatment. I beat them and they went away. They went away with dishonor." For several years in her youth, Nur Jahan stayed in shrines, dressing and behaving as a *pagal*, and singing what she called Baul songs:

> I stayed in the shrine of Shah Ali Baghdadi for about three and a half years. This is in Mirpur. There was a flower tree. But it wasn't alive. In the daytime I stayed under the tree and in the nights in water. Those who saw me [at that time] now say that I used to beg from them. That is, when I was hungry, I asked for food. I was not the least interested in worldly affairs. I visited three hundred and sixty shrines of Bangladesh. I also visited the shrines of one hundred twenty thousand saints, situated in Sonargoan.

Although Nur Jahan suggested in the beginning that she did not remember much from the time she was "*pagal*," in actuality her story is rich with details about her travels, her companions, and her state of mind. What becomes evident in her descriptions is the common polarity of mundane and profane, of "worldly affairs" and "worshipping/singing praises." She was not interested in marriage or material well-being; instead "my only *sadhana* was [focused on] how to get God." Although there is the risk of falling into the temptation to overemphasize a polarity of mundane and profane, often superimposed on cultural phenomena without attention to variations and the significant ways in which worldly affairs and religiosity meld together, what is noteworthy here is that Nur Jahan repeatedly stressed a conflict between these two paths and that she believed that one posed a challenge to the other. First of all, she originally rejected marriage in favor of a life committed to God. She was able to do this as long as she was perceived as being *pagal*; in that state, most men would not want her. Her appearance during that time, as she described it, was certainly not inviting: she had matted hair, carried a machete, and wore dirty black clothes with no jewelry. Furthermore, she claimed that she was able to forget about her own gender: "I lived a different life in my youth. I could not believe that I am a woman. When I stayed with men, I thought that I was a man too. Now I understand everything, good and bad, vice and virtue." These social distinctions were not as important in her previous *pagal* state, living with other *pagals*. She was outside the rules, roles, and judgment of society. As is emphasized by many Sufis, in one's devotion to Allah, gender does not matter. However, when she cleaned up (which she did with the help of a woman who used to visit her in the shrines), Nur Jahan said that she was forced to marry and that it was impossible to carry on as she was without a man. Participating in the world, she recognized the social constructions of good and bad, as she said, and realized her gendered role as a woman.

Having to turn her attention to worldly concerns, such as marriage, caused her much unhappiness:

> I am a very unhappy person, you know. I got married, but not at my will. No, I had no desire [to get married]. People wanted to marry me. Then I had to surrender myself. It was a matter of honor. I was married at the age of twenty-five or thirty. People became crazy seeing me. So I was to be married. Can you get my point? My only desire was to get Allah through songs or music, reciting the name of Him, or through prayers, whatever. This was my only wish: to get Him.

She showed us the scars from being forced to "marry" at knifepoint by a man who desired her.[16] Although she bore him two sons, she said she eventually left him because he refused to let her sing. Her current husband also forbids her from singing (though she did sing for us), and she seemed ambivalent, and somewhat worried, about her staying power in this marriage and in this mundane world:

> Now if I go back to that state of mind—pray/worship, only meditate on Allah and not think of anything else in this world, then I'll go back to being fully absorbed. Then I'll not be able to do anything. That's why I am always keeping my mind on worldly affairs. If my mind goes back to that stage, I will not be normal.

Nur Jahan paused, and with a tone of resignation in her voice, concluded our interview:

> I told you about my [fictive] mother.[17] She died and was buried in Azimpur [Dhaka] . . . She looked after me. She endured many difficulties for me. Sometimes she gave me food, but I refused to take [it]. I threw the food out. But she never beat me. She tried to make me understand. She did up my hair. She helped me in wearing saris also. She kept me inside the house. She did not let me out. Sometimes she told me to look at the mirror: "See how beautiful you are looking?" She wanted to arrange my marriage, but I replied: "I don't know what a marriage is. I don't want to know either." Now I am living a life similar to others. It came from Allah. So I don't blame others for this. Everything comes from Him. Nothing happens without His order. Isn't it?

As is particularly clear in the last few passages of our interview, Nur Jahan's story reveals multiple pulls between living a married life and worshipping God. She never fully reconciles these pulls. As she concluded her interview with us, she stated that if she returned to her worship of Allah, she would no longer be able to remain involved with worldly affairs. Her repeated emphasis on her longing to sing and to worship made me think that this was really a temptation, born at least partly out of her current dissatisfaction. However, she claimed that even this married life came from Allah. Though her life story reveals regret about the situation she is in, she also stated that neither she nor the woman who helped her clean up to the point at which marriage became an option—or a necessity—were ultimately responsible for where she is.

Nur Jahan clearly linked her worship of Allah with losing her ability to stay engaged in the world, and reflecting back on her youth when she was most immersed in worship, she called herself *pagal*. The Baul association with *pagal* has been discussed, but a more nuanced elaboration is needed here in order to understand the myriad meanings and functions of "*pagal*" in Nur Jahan's case. Depending on the context and intention of the speaker, "*pagal*" can be used both to dismiss and legitimize an individual.[18] The few men who reluctantly introduced me to Nur Jahan told me she was a real Baul, and they appeared to have some respect for her. Munibor, whose fictive uncle's house my husband and I stayed in for three weeks, had agreed to help me find Baul women. We spent hours and days engaged in lively and interesting conversation with Munibor, during which time he occasionally discussed Baul women (and his views on women more generally), but he was extremely hesitant about actually introducing me to any. He and many other men I met in Sylhet were keenly worried about my meeting Nur Jahan or any other Baul women, whom they found problematic even if they considered them sincere in their spiritual endeavors.

When I finally had the chance to meet Nur Jahan, he sent along Kasem, a neighboring friend who had been visiting regularly. Kasem sat with us while I interviewed Nur Jahan, and a few times added questions for clarification. Looking back at the transcript of the taped interview, I find those few questions or statements very revealing of how he wanted us to understand Nur Jahan. The first interjection came soon after Nur Jahan began her story, describing, as quoted above, her behavior while living with her parents, staying for hours in the water, and being chained up to keep her in the house. Kasem then offered: "That is, you were *pagal* then?" to which she responded "Yes, I was *pagal.*" Later in the interview when Nur Jahan described how much she loved singing and how she received great happiness from it, Kasem interjected: "This means you love Baul songs very much," clarifying, presumably for our benefit, that the songs she had been singing were Baul songs. The third[19] time I heard him say something was to tell us that she had written several songs of her own. By identifying her as *pagal*, her songs as Baul, and informing us that she had composed songs, Kasem was separating her from other women, many of whom he and Munibor believed had questionable morals, and raising her status by emphasizing her spiritual and musical difference and competence.

One way to understand the significance of her *pagal* behavior and how her status might be raised through these identifications is by considering *majzub*, a concept widespread in the Sufi world. Although I do not recall[20] this term used in my presence when describing any of the Bauls I met, Nur Jahan's descriptions of her behavior most closely resembles that of a *majzub*. A *majzub* is someone who has renounced society and worldly involvement by virtue of her or his being without sense. That is, a *majzub*, like Nur Jahan in her early days, does not know how to live in the world. Why a person so addressed is not merely considered medically crazy has to do with local interpretations of ecstatic devotion or divine possession. Instead of seeing madness as a medical condition, it is understood as reflecting a close vicinity to God. "The *majzub* is one whose speech and actions appear to lack sense because his or her mind has been 'burned' by the closeness to God"

(Ewing 1998:160). She or he may be actively worshipping God as well, or it may be others who interpret a *majzub* as being spiritually gifted or acting religiously; often a person called *majzub* by one person is dismissed as being crazy in a more mundane or mental sense by another person. The Bengali use of *pagal* may also have this divine association, as seen when Bauls are described as being *pagal* / mad with love for the Divine, but it also can be used to degrade someone.[21]

In this case, I think that Kasem's use of the word *pagal* in describing Nur Jahan reflects this localized interpretation of being "burned" by the closeness to God, as is perhaps more clearly recognized in the term *majzub*. That Kasem regarded Nur Jahan's past *pagal* behavior as suggesting a closeness to God is supported by his identification of her as a Baul (in an earlier conversation), a term also used for his close friend and our host Munibor. Unlike most Bauls I met in Sylhet, Munibor is highly educated, intelligent, and charismatic. He is also widely known and respected for having committed the Qur'an to memory. He is the teacher of several students, and many come to listen to him tell moralistic and entertaining stories or discuss the Qur'an and spiritual practices. In contrast, Nur Jahan is not educated, a point she was embarrassed enough about to repeat several times to me. Although Munibor and Kasem expressly stated that they were ashamed and disgusted by the lack of education of many Bangladeshis, particularly Baul women, they appeared to make an exception in Nur Jahan's case. Furthermore, Nur Jahan is clearly a gifted singer. When I heard her sing, which I was able to do one afternoon in the absence of her husband, I felt her deep and soulful voice filling the room with her intense sincerity, silencing those in its path. It seemed that Nur Jahan's ability to sing well and compose her own songs as well as her *pagal* behavior, if interpreted as reflecting a closeness to God, served as grounds to grant her some respect by these scrutinizing men.

I bring up *majzub* also because it sheds light on Nur Jahan's attempt to renounce society. Nur Jahan did not have an institution of renunciation like her Vaishnava Baul counterparts through which she could delve unhindered into her spiritual world. But she did have the cultural constructions of *majzub*, loosely shared by the Bengali *pagal*, which offered her some insulation from worldly concerns. As Nur Jahan emphasized, "Yes, I am *pagal*. But *pagal* about singing. I am telling you the truth. I don't know what other singers do. My only practice was to worship God. I always tried [to figure out], how could I get Allah? I tried to do the things through which I could easily get Him." Kasem and Munibor were not dismissing Nur Jahan by calling her *pagal*, they were giving her a legitimate and, I believe, honorable excuse for her many years living unmarried at shrines and singing praises to Allah.

But Nur Jahan also reveals the difficulties women face when trying to leave society and married life in order to pursue a religious life. Nur Jahan wanted to leave worldly concerns. But as a Muslim in Bangladesh, she did not have the institutional resource of a *sannyasi* tradition that Vaishnava Bauls have with which to affiliate herself (even if only symbolically). Nur Jahan's difficulties also emphasize from another side the significance of the concerns expressed so clearly by Baul women like Madhabi. Being a woman alone in Bengali society *is* difficult. And taking renunciation, even for Bauls who more typically take this step as a couple, puts women in a precarious position vis-à-vis the larger Bengali society.

Conclusion

Although renunciation conjures up lofty and sentimental images for many Bengalis, shared also by a good number of Bauls, I suggest that Baul renunciation actually has more to do with society than it does with the Baul path. There is no essential reason intrinsic to the path of Bauls to take renunciation: it is not a requirement or an established step on the path and, as I discussed, renunciation does not necessarily indicate a dramatic transformation from householder status to ascetic, as is encouraged by many other renouncer traditions.

However, renunciation can be an important way to negotiate one's position in society, and it is in that light that we need to understand these Baul women. Being a Baul woman is in itself a challenge to traditional roles for women, and many Baul women attempt to maintain a respectable status in the community of non-Bauls in which they live. Renunciation further challenges their gendered role—not so much because their life circumstances change, but because other people's perception of them does. But it is also this change in perception that entices some single women to take renunciation because they see the status of renouncer as offering them more than merely being alone. As renouncers they are no longer viewed merely as unmarried women, widows, or wives abandoned by husbands. Significantly, renunciation can serve as a way of providing these women with some semblance of normalcy (through the association with *sannyasi* traditions) while at the same time enabling nonconformist behavior.[22]

To return to the question of what these women are actually renouncing when they take this step alone or with a partner, I suggest that they are not renouncing society at all, but society's expectations of them. As renouncers or *pagals*, they have created for themselves a space in which they can choose how involved they will be with society and family. For most of the women I interviewed, renunciation did not precipitate any severing of social ties, but it did provide them with an excuse for evading societal roles and expectations, and as single women alone they were then able to pursue their musical and/or spiritual path. Thus not only do they gain a certain amount

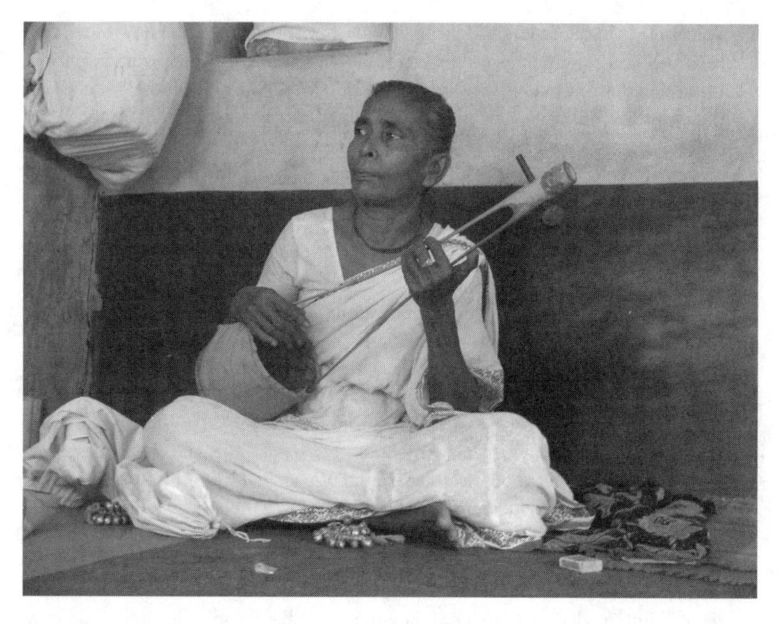

Figure 8.1 Phulmala Dasi Baul, West Bengal, 2005
Source: Photo by Santanu Mitra.

of legitimacy for otherwise nonconformist behavior, but they also loosen society's claim on them (figure 8.1).

Acknowledgments

The research on which this essay is based was funded by Fulbright in 1998–1999 and in 2000 for fieldwork in India and by the American Institute of Bangladesh Studies during 1999 and again in 2000 for fieldwork in Bangladesh. My appreciation goes to the editors of this volume: Ann Grodzins Gold, Sondra Hausner, and Meena Khandelwal for their very helpful comments on this essay.

Notes

1. These terms are often used interchangeably and indicate a formal rite of renunciation, though in my experience *tyagi* and *bairagi* are also sometimes used to convey a sense of detachment without formal renunciation.

2. When recording someone's life story, it is not unusual to wonder how much of it is embellished or altered in the process. My concern with these women's stories is not whether they are factual but rather how these women choose to construct their identity and what they present as their concerns. I agree with Lamb (2001) that telling one's own story is an act of agency, as one reconstructs one's self and history in the process of the telling. Contradictions in different tellings of one's story can also serve to reveal issues of contention and a process of negotiating with conflicting ideals and expectations.
3. These regions are important to the identity and character of the Bauls discussed here. Although it is beyond the scope of this essay to address fully, being a Baul has different connotations and implications in each of these regions, though there are also very important similarities.
4. Baul identity as well as Baul renunciation is the subject of much debate among scholars, lay people, and Bauls themselves. What is striking in all this discourse is that there is no consensus about what constitutes a "real" Baul, yet many make claims of authenticity. Openshaw (2002) and Urban (1999) take up various issues of this debate in their thoughtful work, and elsewhere (2005) I have also considered questions about Baul identity as they pertain to women who either call themselves Bauls or are called so by others. In this essay I do not engage in the debate about Baul authenticity but rather choose to privilege those who identify themselves as Bauls. My objective therefore is not to decide if they meet particular criteria of "Baul" as all the research has shown me that Baul is actually a very fluid term that is applied in different ways to different people and in various regions. In fact, what should be of interest in this essay to those with knowledge about "Bauls" is that the term is indeed used to describe some very different people.
5. Literally translated, *madhukari* means gathering honey; the act of begging from door to door is likened to a bee gathering honey from flower to flower. The bee (*madhukar*), like the wandering mendicant, is seen as taking a little nectar from many flowers, never a lot from one place.
6. Furthermore, as pointed out by Openshaw (2002:140–151) and also corroborated by my fieldwork, the guru who gives *bhek* is not necessarily seen as the most important guru to Bauls. Openshaw argues that because a Baul may have multiple gurus and value the *shiksha* guru who teaches them songs and/or *sadhana* over the other gurus, the role of the *bhek* guru as well as the status of *bhek* is effectively devalued.
7. Among Bauls, marriages are often marked by exchanging garlands in the presence of a guru, and this can take place for second or third marriages as well, though it is less likely. More common is a couple simply deciding to be together and assuming a state of marriage. I suspect that this is the case with Kangalini, but any further interrogation would, I believe, have been fruitless and essentially inconsequential. What is important is that Kangalini changed her status from Hindu to Muslim and from single to married and as a result she says she experienced much greater acceptance in the Muslim community in Bangladesh.
8. Kangalini is referring to the dark color of her skin. In South Asia, dark skin color is considered less desirable, and a young woman who is dark may have difficulties getting married and will sometimes receive ill treatment because of her skin color.

9. Kangalini's description of a woman's duties sounds remarkably Hindu. Part of the reason for this is that she herself was raised a Hindu. Furthermore, in South Asia many Hindu and Muslim views mix and blend. This is certainly the case in Bangladesh.
10. Madhabi often spoke using fragmented sentences, which makes for some confusing reading. Instead of retaining the original rhythm of her speech in translation here, I have decided to paraphrase her story in this instance in order to focus on the general description of her concerns. As a result, I have filled in a few words and phrases that were omitted but implied in the context of her dialogue. Other quotes from Madhabi in this essay are direct translations.
11. Some Bauls claim that a place can only be considered an ashram if there is a grave (*samadhi*) there. Unlike most traditional Hindu householders, Bauls bury their dead, a practice shared by some other ascetics and renouncers as well as Muslims.
12. Wadley (1995) demonstrates how a widow's position depends on her relationship to other males such as sons, fathers-in-law, brothers-in-law, and brothers, and her natal and affinal families make decisions about which of her roles in relation to those men will dominate. In Madhabi's case, her primary role after her husband left was as a mother to her only daughter, and her parents helped her fulfill that role by assisting in the daughter's marriage arrangements. It is this role that is still her concern, though since most of her responsibilities in that role have been met, she is also attempting to secure an independent position as a Baul singer. See also Lamb (2000, 2001) for an excellent study of widows in West Bengal. She found that when a widow has a child or children, it is her role as mother, not wife, that emerges most often in her self-representations (2001).
13. In Bengal, *maya* is commonly used to mean both illusion and the attachments one has with loved ones such as family, children, and neighbors. See Lamb (2000) for more on this understanding and use of the term *maya*. More standard views of renunciation suggest that one should loosen one's attachments to others in order to break from the cycle of rebirth. Although Bauls often do not actually loosen such ties, Jaya draws on these more expected views of renunciation when articulating her decision not to renounce because those views are the ones shared by her neighbors.
14. While this eagerness to pursue the path alone may lead some to question their Baul identity, both Nur Jahan and Rajani find resonance with the label that others also use in referring to them. Furthermore, their perspectives on being a wife and taking renunciation contribute not only to our present discussion of what it means for Baul women to take renunciation, but also similar discussions with Baul women in which they engaged at the times I interviewed them. Finally, Nur Jahan represents another interesting layer to Baul identity that is evident in parts of Bangladesh, particularly in areas at some distance from regions where followers of Lalan Shah are found. In my research in Sylhet, I found that the term "Baul" was used to describe certain people viewed to be outside of normative society. Often they were also artists who expressed their longing for the Divine in song and would behave in unconventional ways. But unlike followers of Lalan and most of my other informants in West Bengal, they did not, to my knowledge, engage in sexual ritual practices.

15. James M. Wilce's (1998) discussion of Bangladeshis dismissing someone or rendering them powerless by calling them *pagal* may provide some insight here. Although it does not appear that Nur Jahan nor those who introduced me to her dismissed her religiosity, her reference to herself as *pagal* in the past appears to serve as an excuse for behavior that may be perceived as unacceptable.
16. Although Nur Jahan used the word to marry (*biye*), the implication appears to be that he raped her.
17. This woman, whom Nur Jahan calls her mother, used to visit her at the shrines. Nur Jahan credits her with helping her clean up and become engaged with worldly affairs. As Nur Jahan had begun cleaning up, she said she was able to regain her ability to speak after many years (presumably she had no difficulties singing). It was then that her "mother" asked her what she wanted, to which she responded that she wanted to sing professionally. Nur Jahan says she became the first female singer in Sylhet and the third in Dhaka.
18. See James M. Wilce (1998) for an interesting discussion on *pagal* and legitimacy in the Bangladesh context. See McDaniel (1989) for more on divine madness in West Bengal.
19. Kasem said something two other times before this, but they were inaudible. Nur Jahan sent him away soon after this.
20. At the time of fieldwork, I was not familiar with the concept *majzub*. If it had been used at all by any of the people I had been interviewing, it is likely they also added, perhaps for my benefit, an additional word such as *pagal*, *udasi* (free from worldly concerns), or *marami* (mystic), all of which were very common in addition to *baul* in its descriptive sense.
21. For instance, Jim Wilce (2002) shows that the term *pagal* was used by conservative Muslims to describe and restrain an old man who preferred to sing aloud about his love for Allah, a practice they disapproved of.
22. This applies to Baul couples too. For instance, renunciation can give them a cover of some legitimacy, or allow what is socially viewed to be an unsuitable couple (e.g., Muslim and Hindu, or Hindus of different castes) to live together.

Bibliography

Abbas, Shemeem Burney. 2002. *The Female Voice in Sufi Ritual: Devotional Practices of Pakistan and India*. Austin: University of Texas Press.

Abu-Lughod, Lila. 1990. The Romance of Resistance: Tracing Transformations of Power Through Bedouin Women. *American Ethnologist* 17:41–55.

Ahearn, Laura M. 2001. Language and Agency. *Annual Review of Anthropology* 30:109–137.

Burghart, Richard. 1978. The Founding of the Ramanandi Sect. *Ethnohistory* 25(2):121–139.

———. 1983. Renunciation in the Religious Traditions of South Asia. *Man* 18:635–653.

Caudhuri, Abul Ahsan. 1997. Bangladesher baulder calacitra. In *Dhruvapad: Banglar baul phakir*. S. Cakrabarti, ed., pp. 131–137. Nadiya, West Bengal: Krishnanagar.

Dasgupta, Shashibhusan. 1962[1946]. *Obscure Religious Cults*. Calcutta: Firma K.L. Mukhopadhyay Publishers.

De, Sushil Kumar. 1942. *Early History of the Vaisnava Faith and Movement in Bengal*. Calcutta: General Printers and Publishers.

Denton, Lynn Teskey. 1991. Varieties of Hindu Female Asceticism. In *Roles and Rituals for Hindu Women*. J. Leslie, ed., pp. 211–231. London: Pinton Publishers.

Ernst, Carl W. 1999. *Teachings of Sufism*. Boston: Shambhala Publications.

Ewing, Katherine P. 1998. A *majzub* and his Mother: The Place of Sainthood in a Family's Emotional Memory. In *Embodying Charisma: Modernity, Locality and the Performance of Emotion in Sufi Cults*. P. Werbner and H. Basu, eds., pp. 160–183. New York: Routledge.

Frembgen, Jurgen Wasim. 1998. The *majzub* Mama Ji Sarkar: "A Friend of God Moves from One House to Another." In *Embodying Charisma: Modernity, Locality and the Performance of Emotion in Sufi Cults*. P. Werbner and H. Basu, eds., pp. 140–159. New York: Routledge.

Gal, Susan. 1992. Between Speech and Silence: The Problematics of Research on Language and Gender. In *Gender at the Crossroads of Knowledge: Feminist Anthropology in the Postmodern Era*. M.D. Leonardo, ed., pp. 175–203. Berkeley: University of California Press.

Ghurye, G.S. 1953. *Indian Sadhus*. Bombay: Popular Book Depot.

Gold, Ann Grodzins. 1992. *A Carnival of Parting: The Tales of King Bhartari and King Gopi Chand as Sung and Told by Madhu Natisar Nath of Ghatiyali, Rajasthan*. Berkeley: University of California Press.

Hayes, Glen Alexander. 2000. The Churning of Controversy: Vaisnava Sahajiya Appropriations of Gaudiya Vaisnavism. *Journal of Vaisnava Studies* 8(1):77–90.

Khandelwal, Meena. 1996. Walking a Tightrope: Saintliness, Gender, and Power in an Ethnographic Encounter. *Anthropology and Humanism* 21(2):111–134.

———. 1997. Ungendered Atma, Masculine Virility and Feminine Compassion: Ambiguities in Renunciant Discourses on Gender. *Contributions to Indian Sociology* 13(1):79–107.

———. 2004. *Women in Ochre Robes: Gendering Hindu Renunciation*. Albany: State University of New York Press.

Knight, Lisa I. 2005. Negotiated Identities, Engendered Lives: Baul Women in West Bengal and Bangladesh. PhD dissertation, Department of Anthropology, Syracuse University.

Lamb, Sarah. 2000. *White Saris and Sweet Mangoes: Aging, Gender, and Body in North India*. Berkeley: University of California Press.

———. 2001. Being a Widow and Other Life Stories: The Interplay between Lives and Words. *Anthropology and Humanism* 26(1):16–34.

McDaniel, June. 1989. *The Madness of the Saints: Ecstatic Religion in Bengal*. Chicago: University of Chicago Press.

O'Connell, Joseph T. 1982. Jati-Vaisnavas of Bengal: "Subcaste" (Jati) without "Caste" (Varna). *Journal of Asian and African Studies* 17:189–207.

———. 1990. Do Bhakti Movements Change Hindu Social Structures? The Case of Caitanya's Vaisnavas in Bengal. In *Boeings and Bullock-Carts: Studies in Change and Continuity in Indian Civilization*. B.L. Smith, ed., pp. 39–63. Delhi: Chanakya Publications.

Ojha, Catherine. 1981. Feminine Asceticism in Hinduism: Its Tradition and Present Condition. *Man in India* 61:254–285.

———. 1988. Outside the Norms: Ascetics in Hindu Society. *Economic and Political Weekly*: WS 34–136.

Openshaw, Jeanne. 1994. "Bauls" of West Bengal: With Special Reference to Raj Khyapa and His Followers. PhD dissertation, Department of Anthropology, School of Oriental and African Studies, University of London.

———. 2002. *Seeking Bauls of Bengal*. Cambridge: Cambridge University Press.

Ray, Benoy Gopal. 1965. *Religious Movements in Modern Bengal*. Santiniketan: Visva-Bharati Research Publications.

Salomon, Carol. 1991. The Cosmogonic Riddles of Lalan Fakir. In *Gender, Genre and Power in South Asian Expressive Traditions*. A. Appadurai, F.J. Korom, and M. Mills, eds., pp. 267–304. Philadelphia: University of Pennsylvania Press.

———. 1995. Baul Songs. In *Religions of India in Practice*. J. Donald S. Lopez, ed., pp. 187–207. Princeton: Princeton University Press.

Schimmel, Annemarie. 1996. *My Soul Is a Woman: The Feminine in Islam*. S.H. Ray, trans. New York: Continuum Publishing.

Smith, Margaret. 1984[1928]. *Rabi'a the Mystic and Her Fellow-Saints in Islam*. Cambridge: Cambridge University Press.

Tripathi, B.D. 1978. *Sadhus of India: The Sociological View*. Bombay: Popular Prakashan.

Urban, Hugh B. 1999. The Politics of Madness: The Construction and Manipulation of the "Baul" Image in Modern Bengal. *South Asia: Journal of South Asian Studies* 27(1):13–46.

Vallely, Anne. 2002. *Guardians of the Transcendent: An Ethnography of a Jain Ascetic Community*. Toronto: University of Toronto Press.

van der Veer, Peter. 1988. *Gods on Earth: The Management of Religious Experience and Identity in a North Indian Pilgrimage Centre*. London: Athlone Press.

Wadley, Susan S. 1995. No Longer a Wife: Widows in Rural North India. In *From the Margins of Hindu Marriage: Essay on gender, Religion, and Culture*. Lindsey Harlan and Paul B. Courtright, eds., pp. 92–118. New York: Oxford University Press.

Wilce, James M. 1998. *Eloquence in Trouble: The Poetics and Politics of Complaint in Rural Bangladesh*. New York: Oxford University Press.

Wilce, Jim. 2002. Tunes Rising from the Soul and Other Narcissistic Prayers: Contested Realms in Bangladesh. In *Everyday Life in South Asia*. Diane P. Mines and Sarah Lamb, eds., pp. 289–302. Bloomington: Indiana University Press.

Chapter 9

These Hands Are Not for Henna

Anne Vallely

Introduction

The events of the first evening that I arrived in Ladnun were uncharacteristic of all the rest: raucous laughter, singing, and the rhythms of clanking bangles rang through the air, breaking the silence of the monastery. The soft chants of the ascetics' prayers were drowned out by the revelry of a wedding party a few blocks away.

Although I had traveled far to live among women renouncers of the Terapanthi Jain ascetic community, curiosity got the better of me: I ventured out of the grounds to watch the merrymaking in the open courtyard across the street. Women in ornate saris danced in circles, pounding rhythms into the sand with jeweled and amber-stained feet. Their hands with beautiful henna patterns moved deftly to the music, causing the bangles on their wrists to jingle in sync.

Moments later I returned to the grounds of the monastery. The nuns were sitting quietly after saying their evening prayers. With the sounds of celebration still around us, I sat down before one young nun to describe what I had just seen. She waited a little before responding, and when she did, all she said was, "In the world, you care about the body. Here we care about our souls." This was a message I would hear many times during the year I lived among the Jain nuns.

Dressed in stark white, they radiate simplicity and purity. With their unadorned bodies they proclaim a radical message repudiating worldly life:

> These hands are not for henna, nor are they for washing, cooking or caring for others. These hands are tools of non-violence; careful instruments of deliberation. And these feet are not for scurrying after little ones, shuffling on errands in sandals, not for decorating or dancing at weddings. They are for steady steps along a deliberate and pious path. My body is my instrument; it is not me. I decide when, or whether to feed it; when, or whether to rest it; And I will decide when it serves me no more, to end my connection with it forever.

In this essay, I explore the female religious body in the Jain renunciatory tradition, because—perhaps more than in any other way—Jain women narrate their own experiences of asceticism with reference to the body. The path from lay to ascetic life and spiritual progress along the ascetic path, largely entails an ever-increasing sensitivity to bodily otherness—to body objectification and symbolism. Through ascetic discipline, the "worldly" comes to be identified by the renouncers in terms of their former socialized bodies, as well as in terms of the socialized bodies of lay women around them. Renunciation is a momentous process of undoing, occurring at the level of embodied experience.

Outwardly, at the symbolic, cultural level, female renunciation within the Jain tradition is rarely seen as "radical." It has not garnered the same level of admiration as has its male counterpart. Male renunciation, for reasons discussed below, is better able to represent "rupture" with lay worldly life. And symbolically, this rupture is the necessary ground upon which heroism in the Jain tradition is built. As a lived practice, however, female renunciation is indeed radical: It results in a profound shift in the relationship between self and body. Renunciation allows for the objectification or "othering" of the body—a crucial aim of Jain spiritual practice, but one that is difficult for lay women to achieve.

To forget oneself in daily activities is a form of surrender to the dictates of the body and social life. The body hungers; it cries out for affection, comfort, and sustenance. In its vanity it seeks splendor and approval, and anxiously masquerades its decay. Worldly life is a spectacle of fleeting corporeal flesh, filling up hungers whilst creating appetites anew.

The self is a small voice, muffled and buried beneath the body's demands. It is vulnerable and bewildered in the chaos of worldly life. It forgets that it does not belong to the body. Deluded, it mistakes itself for the flesh. Duped, it identifies with social demands and social roles, and seeks its affirmation in this physical realm. Swept along with the dictates of social, bodily life, it is almost invisible. Renunciation changes all this: in turning away from society, it renders the body mute. Without a voice of its own, the body comes to reflect its "keeper," the soul, or true self.

The Ascetic Imperative

Jainism teaches that every living being has a perfect, eternal soul (*jiv*), and that all souls deserve compassion. All are in a state of spiritual impurity, bound to earthly existence because of karmic shackles. The universe is filled with pitiable souls, bound to remain in physical existence because of ignorance of the basic causes of their captivity. Teachings are based on a division of all existing things into two main classes: *jiv* (soul or that which is sentient) and *ajiv* (an insentient material component that is attracted to the soul). *Ajiv* attaches to *jiv*, obscuring the true and omniscient nature of the soul, and causing it to be cast into an endless succession of births and deaths.

Liberation is a state of bliss and omniscience, but it can occur only when the soul frees itself from its ruinous captivity. This is not a simple task: the connection between *jiv* and *ajiv* is beginningless, and is sustained by passions (*kashaya*)—attachment, aversion, greed, and envy. It can only be severed through sustained and rigorous efforts at detachment from worldly life. Jain moral practice is characterized by the cultivation of self-restraint—the essence of detachment from the material world of karma—and by a deeply held compassion for all living beings. Jain ethics are enshrined in the *mahavratas*—the five "great vows" that work to "fence in" or limit worldly entanglement. Nonviolence (*ahimsa*) is the first and most fundamental vow. It demands nonviolence in thought, speech, and action, and it is believed to be the foundation upon which the remaining four derive, namely Truthfulness, Nonstealing, Sexual restraint, and Nonpossession.

Fundamental to Jainism is the dualism between the body and self. The two are distinct and antithetical, corresponding with the metaphysical dualism of *jiv* and *ajiv*. In contrast to contemporary Western thought, dualism is considered neither a curse nor a condition to overcome, but a reality that is too often forgotten in the clamor of worldly life. Rather than seeking ways to overcome dualistic thought and experience, Jainism seeks to increase our awareness of it.

Jain dualism is every bit as encompassing as that of the Western-Cartesian variant, but of course, it is distinct from it. For one, Jainism does not treat the human soul as alone or unique in a soulless universe. Instead, the universe is animated with living beings, all of which have a soul. In addition, the soul-body relationship in Jainism is far more indexical than Descartes would have allowed: the body is not an arbitrary vessel sheltering the soul but mirrors the state of the soul at the time of its incarnation. The body is the precise expression of one's karmic "baggage": its appearance, health, intelligence, and status are all outcomes of its past. It is believed that the soul absorbs the zygote formed by the blood of the mother and semen of the father in the womb, and creates the physical body or fetus (Tatia 1994:52). Furthermore, the soul is not confined to a fixed location, such as the brain or heart, but permeates every dimension of the body, corresponding with the body's shape. A property of the soul is its ability to assume whatever bodily form it inhabits.[1]

Despite this indexical relationship, the body and soul are fundamentally distinct entities. The goal of the Jain path is to disentangle permanently the soul/matter union. And while the fusion of soul and matter is without a beginning, the soul's true nature is that of a disembodied, omniscient, and blissful state.

The metaphysical categories of *jiv* and *ajiv* are recognized in the experience of the self and body as separate entities. The body is the expression of one's karmic burden, ensnaring and even altering the soul, but fundamentally separate from it.

One's body is the limit and boundary of one's control. Beyond it, and at all times pressing against it, is a world teeming with life. There is no empty space for the body to move through, so one must tread with care to avoid harming others. Unlike Descartes' view of the universe as a soulless cavity, made special only through the human presence, the universe for Jains pulsates with significant life. It is filled with a great number of souls, all of which are essentially identical, but differently embodied. These differences in bodily forms are important and instructive in that they convey information about different states of worldly bondage. The human form, for instance, is an accomplishment in and of itself; a privileged incarnation, achieved through past meritorious behavior. In addition, bodies can be "read" to reveal different levels of self-awareness. For instance, air-earth-water-and-fire beings are so heavily oppressed by their karmic burden that their level of self-awareness is virtually nonexistent. Human beings, at the opposite end of the spectrum, have the greatest degree of self-awareness, and it is only through the human body that *moksha* (enlightenment) can be achieved.

To be born in human form signifies the accumulation of auspicious karma in a previous birth. And notwithstanding a few spectacular Jain fables to the contrary, it is only humans who have enough knowledge to realize that they are in a state of worldly bondage, and enough courage to renounce worldly life.

All souls are in a state of karmic bondage, trapped in a worldly quagmire of their own doing. Each incarnation or body is, in essence, symptomatic of past karmic debts, paid and owing. Yet, all souls, from the simplest one-sensed *nigodas* to five-sensed animals, including human beings, yearn for release. And to obstruct or destroy another soul's journey, no matter how humble its life form, is the gravest of transgressions, and inevitably results in karmic consequences.

The physical world for Jains is a moral theater. It is the foundation upon which the ethical self is created or doomed. Not only proper relations with human others, but interactions with all living beings constitute an ethical way of being. To fully understand the implication of this is rare. Most living beings are too adrift, karmically speaking, to care. But for those who are truly awake to the violence inherent in life, there is little choice but to repudiate it thoroughly.

Paul Dundas captures the essence of the Jain worldview, out of which the ethical imperative of asceticism arises, as it is presented in the Acaranga Sutra.[2] He writes:

> The world is characterized by ignorance, suffering and pain caused by action. . . . True understanding embodies itself in non-violence through an awareness that all living creatures, including oneself, do not wish to suffer in any way. . . . As a broad ethical principle, this is fairly unexceptional and has to be fitted into a further series of conceptions: action, whether done, caused or condoned by oneself, brings about rebirth . . . and the world is in a state of suffering caused by actions of ignorant people . . . who do not know that they are surrounded by life-forms which exist in earth, water, air and fire, a true understanding of which can be gained from the teaching of Mahavira. (1992:36–37)

Renunciation is a logical response to the teachings of Jainism. Those who choose the ascetic path are believed to have (and believe themselves to have) awoken to truth. So sensitive are they to the sufferings and meaninglessness of worldly life that they feel compelled to wholly reject it. They believe that it is only through renouncing worldly existence that one can hope to lead a truly nonviolent life. Lay life, no matter how virtuous, is unavoidably mired in the mundane violence of the "ignorant." The most banal of acts such as eating, talking, sleeping, bathing, and walking are inextricably caught up in violence.

Women, the Body, and Asceticism

Since the fourth century BCE the Jain community has been divided into two major branches: the Svetambars (whose renouncers wear white robes) and the Digambars (whose ascetics are "sky clad" or naked) (Dundas 1992:43) (figure 9.1). There are several areas of contention between them, but one of the key disputes centers on the issue of female spirituality.

The debate concerns the question of whether or not *moksha* (liberation) can be attained following a life in a female body (Balbir 1994; Jaini 1991).

Figure 9.1 Svetambar nun, Jain temple Nakodu, Rajasthan

Source: Photo by *Jain Spirit Magazine* www.jainspirit.co.

The Digambars argue that since practicing nudity, an essential ascetic practice, is not a realistic prospect for women, liberation is not possible following a life in a female body. They argue that a woman would have to be reborn as a man before liberation could occur. The Svetambars, who do not consider garments to be an obstacle to liberation, argue that women can attain salvation (Balbir 1994; Jaini 1991; Banks 1986; Shanta 1985). The practical consequence of this doctrinal debate is that fully fledged female ascetic orders

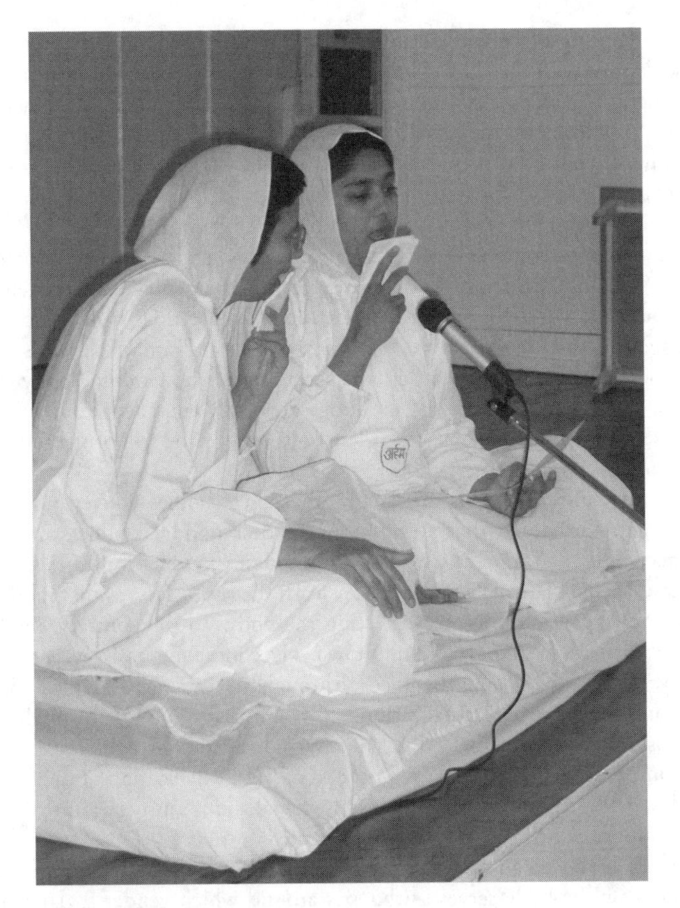

Figure 9.2 Terapanthi samanis from Jain Vishwa Bharati, Rajasthan

Source: Photo by *Jain Spirit Magazine* www.jainspirit.co.

exist only among the Svetambar sect. It admits women to full monastic vows. Among the Digambar, women are not permitted to take full monastic vows and therefore can attain only the quasi-ascetic status of *aryika* meaning "noble woman" (Babb 1996; Jaini 1991; Shanta 1985:483–517).

The Terapanthi Jains, with whom most of my work has been conducted, and who serve here as the template for my discussion of female asceticism, are a branch of the Svetambar sect (figure 9.2). With nearly six hundred nuns, they have the largest order of female ascetics under a single *acharya* (leader).

Although the Svetambar and Digambar are often depicted as opposites with respect to their views on female religiosity, the ideas they hold in common are as important as those on which they differ. Both hold the same understanding of female nature as flawed, because of women's presumed associations with sexuality. The force and persistence of these ideas inform practices of gender socialization and form the basis of a religious imagination, which problematizes the female form as a cultural symbol of renunciation.

But, of course, women today do not pursue the ascetic path in order to be cultural icons. They pursue the ascetic path for a variety of personal reasons, of which faith is usually predominant.[3] Consequently, their failure to serve as cultural symbols—of significance to those concerned with strategies of female subordination—has little effect on their religious experiences. Renouncing worldly life to pursue the life of an ascetic is truly a momentous event, one that alters the way they understand and experience their embodied selves. It forces an alternate reading of the feminine body. The body, so long accustomed to "speaking" a worldly language of nurturance and sexuality, must now become silent so that only the will of the self/soul is heard.

From childhood, girls are taught that their bodies convey powerful messages: they learn that "proper" or modest bodily conduct indicates respectability and morality, whereas "immodest" conduct indicates impropriety, usually of a sexual nature. They learn the subtle lessons that modesty does not mean ignoring or neglecting the body. To the contrary, modest behavior is really the art of simultaneously concealing and revealing the body. The body is still the focus, through its "absence." Frigga Haug explores how female socialization, cross-culturally, occurs through the imposition of a set of behavioral norms and restrictions, which call forth and imply sexuality. She calls it a process of the "sexualization of the female body" (1987). Through this process, the female body's association with sexuality appears intrinsic and insurmountable.

Leela Dube, in an essay on the construction of gender in India, writes about how the body serves as the axis around which gender socialization occurs. Although her focus is on the socialization of Hindu girls, the same patterns can be observed within the Jain tradition.

Considerable importance is attached to the way a girl carries herself, the way she sits, stands and talks, and interacts with others. A girl should walk with soft steps: so soft that they are barely audible to others. Taking long strides denotes masculinity. Girls are often rebuked for jumping, running, rushing to a place, and hopping. These movements are considered part of masculine behavior, unbecoming to a female; however, the logic of the management of a girl's sexuality also defines them as unfeminine; they can bring the contours of the body into greater prominence and attract people's attention. A girl has to be careful about her posture. She should not sit cross-legged or with her legs wide apart. Keeping one's knees close together while sitting, standing, or sleeping is "decent" and indicates a sense of shame and modesty. "Don't stand like a man" is a common rebuke to make a girl aware of the demands of femininity (1988:16).

The female body is socialized to be a medium of expression, but the language it "speaks" has a limited vocabulary. Its capacity for symbolism is restricted to corporal idioms of nurturance and sexuality (which, of course, are worldly idioms). In social life, the female body speaks the language of modesty/sexuality, of fertility/barrenness, of nurturing/abandonment, of faithfulness/infidelity. The capacity of the female body to signify transcendent values appears constrained; the image of the feminine does not typically represent cultural ideals of nonviolence, restraint, dispassion, courage, rationality, and so on. Female symbolism remains centered on values connected with the body and sexuality, and associated with the social roles of daughter, wife, and mother.

Even the religious practices of lay women (prayer, fasting, meditation, temple life, etc.), do not remove them from their association with sexuality. Instead, religious observance is a fundamental marker of sexual purity and female honor. So, while religious observance establishes good character, it does not allow women to transcend associations with sexuality. Josephine Reynell writes about the way religion is the primary means through which female honor in Jainism is privately and publicly demonstrated:

> Men must also demonstrate a degree of moral uprightness, but this is accomplished less through actual behaviour and more through using their wealth in a particular way. Consequently, there is less scrutiny of their actual daily behaviour, whereas for a woman the converse is true. They continuously demonstrate their honour, and that of their family, through their behaviour, which must be impeccable. In particular, it is through their religious activities that they express their moral worth to the community. (1985:162–163)

The religious practices that laywomen observe for their own personal spiritual development are inseparable from the messages these practices convey about female sexuality and honor.

A laywoman's body speaks this worldly language irrespective of her individual traits, ambitions, or accomplishments. In addition, female social roles are likewise thoroughly "corporal." They are centered on caring for others: cooking, feeding, comforting, cleaning, and so on—all of which are problematic from the perspective of Jain ideals of nonviolence.

It is important to note here that Jain renouncers consider all forms of social service (*seva*) to be worldly activities, unbefitting the spiritual path. Nevertheless, service to other renouncers *is* acceptable and, in fact, constitutes an important part of their spiritual discipline. In addition to the *seva* that renouncers provide each other on a regular and spontaneous basis, nuns perform many traditional gender-based tasks (such as the sewing of garments) for monks. Meena Khandelwal's work on Hindu renouncers (in this volume) highlights the importance of gender, as well as the gendered practice of *seva*, in the lives of many female renouncers. Her arguments are germane for the Jain renouncer tradition as well, where *seva*—although problematized in its worldly expressions—is an institutionalized and gender-based practice within monastic life.

The *seva* that laywomen perform is differently conceptualized. It is a worldly practice, and therefore an obstacle to spiritual progress. The preparation of food entails violence by "killing" plants and water, for instance. And, the emotional ties of family life, though gratifying and beneficial in worldly life, lead to the accumulation of karma, which ultimately impedes one's spiritual path. Roles and expectations enmesh women in worldly pursuits, leaving them little room to cultivate bodily detachment. For most, silencing the body so that the self can speak is an arduous and lifelong challenge.

So intertwined is she in this bodily language that to excise herself from it entails a radical process through which her body loses its ability to express itself. In some cultures, the onset of old age has the effect of disconnecting women from the sexualized body. Postmenopausal bodies, for instance, when viewed as being outside the realm of sexuality, become "neutral vessels," and lose their ability to define women. Free to define themselves, women commonly experience increased prestige and power (see Kerns and Brown 1992). Renunciation has a similar effect, but, unlike aging, it is a deliberate and abrupt end to the woman-as-body association (see Shneiderman, this volume, for an alternate example of this process among female Bon renouncers (*chomo*) of the Nepal Himalayas). Renunciation causes an end to the body's expressiveness. Messages conveyed through the body are no longer of, or about, the body. Unable to communicate its own, corporal messages, it now serves as a medium for the expression of the self.

When a young woman (or girl) of Jain faith aspires to renounce worldly life and pursue the religious path, she must demonstrate her disenchantment with the world in general, and with her own body in particular. Since our bodies are the most cherished of our attachments, they are the most intimate and condensed expressions of the worldly. Confusing the desires of the body with those of the true self leads to bewilderment, which is the primary source binding karma and causing rebirth. Genuine detachment from the body is an especially difficult task, but when achieved, it signifies a true hero—a true Jain. The name "Jain" derives from the Sanskrit word *jina*, meaning "conqueror." The *sadhvis* and *munis* are heroes who have "conquered" themselves—that is, conquered their bodily passions.

Othering the Body

Female aspirants demonstrate their preparedness for the ascetic life through a fascinating process in which they acquire the attitude, language, and demeanor of worldly detachment (*vairagya*). The single most important method is through "othering" the body. They cultivate an attitude of indifference to the well-being and appearance of their bodies, and master the discourse of bodily objectification.

When a Jain girl aspires to renounce worldly life and pursue the religious path, she is not in unknown territory, nor is she outside the feminine norm: women renouncers have a long and illustrious history within the Jain tradition, outnumbering their male counterparts, by at least two to one, since the time of Mahavir (Balbir 1994). Nevertheless, she must first persuade her family that her ascetic goals are sincere. This is usually a difficult task that takes considerable time. Despite the honor of having a child pursue the ascetic path, families are typically reluctant to "lose" a child (especially a male child).

Aspirants, therefore, must convince their families that they have had a spiritual awakening, and that worldly pleasures are now meaningless to them. In Jain terms, they must establish themselves as true *vairagini*—one who is possessed of *vairagya* (worldly indifference). On the basis of her work with Jain nuns, N. Shanta, describes this demeanor as follows:

> [*Vairagya*] is a word used frequently by the *sadhvis* in their conversations, lectures, writings and biographies. It expresses the very foundation of their lives, their motivations and reasons for renunciation. *Vairagya* has a both a negative and positive connotation. In its negative sense, *vairagya* suggests a basic indifference toward all of life's pleasures, honors, material wealth, well-being, family relations, friendships etc. In its positive connotation, it asserts that the renunciation of all of possessions, all attachments, everything of worldly life—a life, according to Jain scriptures, that causes one to sink ever deeper into *samsar*—is undertaken for a singular imperative: the knowledge and realization of one's soul [*atman*]. (1985:343; translated by author)

Fasting is the most popular practice of "othering" to demonstrate detachment. Since eating is an activity that sustains our physical selves, it constitutes a link between the *jiv* and *ajiv*, and binds us to *samsar* (worldly existence). It follows that efforts to defy or sever this link take the form of fasting. From about the age of ten, fasting is a common practice for laywomen and girls. Of course, it remains a central practice throughout one's ascetic career. However, for laywomen, it does not necessarily imply worldly detachment. Instead it symbolizes both religious devotion and female honor. Reynell writes:

> [I]n terms of regular temple going, the performance of samayik and pratikraman, the attendance of preachings and pujas and the observance of food restrictions, women are considerably more assiduous and regular in their practice than men. Fasting in particular seems to have become a female sphere of influence and most of the orthodox Jain women undertake quite long and complex fasts, regularly, which gains them both status and public admiration from the rest of the community. (1984:28)

But for girls who wish to prove the sincerity of their ascetic aspirations, the fasts tend to be longer, more frequent, and more difficult than the average, making the link with worldly detachment inescapable. While their professed aim in fasting is to destroy or "burn off" old karma (a process called *nirjara*), equally important is the public display of their self-control. In denying the body its desires, the ascetic aspirants display the supremacy of their will (self/soul) over the flesh.

If they are successful in convincing their families, girls of the Terapanthi Jain community are enrolled in a boarding school called the Parmarthik Shikshan Sanstha (PSS). The school was founded in 1948 by Acharya Tulsi to allow girls to undergo a preparatory training for full-fledged monastic life. The rationale for the school is that aspirants should "be exposed to austerities" in stages and provided with opportunities to understand the difficult path that they are considering. The single criterion of admission is that the aspirant should show positive evidence of her desire to attain the state of final liberation from worldly bondage (Bhatnagar 1985:82). And although her desire to renounce may be tremendous and even overwhelming, her renunciation will follow a definite course.

It should be noted that a difference exists in the process through which young women and men (or girls and boys, as the case may be) come to renounce the world. Their training differs considerably because of the relative small number of male candidates "desirous of emancipation." Young men typically have much shorter training periods, usually gaining entrance into the monkhood within a year. The practical consequence of these arrangements is that young women are far better prepared for ascetic life than are young men at the time of *diksha* (initiation). Yet, this has the ironic effect of undermining the glory of female asceticism. Renunciation is paradigmatically a bold and courageous act; it is an occasion of high drama in which all ties with social life are ostensibly severed. Although the *diksha* ceremony itself attempts to recreate the drama, the fact that women go through a long-term training program, where they essentially "learn" to renounce, undermines the impact of renunciation at the public level.

At the outset of this essay, I claimed that the path from lay to ascetic life entails an ever-increasing sensitivity to bodily otherness and to body objectification. I imagine this to be true for female and male Jain renouncers, irrespective of their sectarian background. For those who aspire to renounce social life, the ability to control their desires, and to cultivate indifference toward their bodies and worldly life, cannot be an instantaneous achievement. For most, it is likely a long-term development that happens imperceptibly until that time when it becomes apparent. What makes the PSS especially interesting is that, because it is a school devoted expressly to this goal, it can be viewed as a kind of laboratory for the process. There are at least two significant features of this process that are observable within the PSS setting: (1) Body otherness is achieved by creating an ever-greater distance between themselves and householders (who are seen as governed by bodily, carnal appetites; (2) Body otherness is attained through acquiring an attitude of disenchantment that comes to frame their experiences of their own bodies as neutral, desexualized tools.

The Carnal Other

The distinction between the worldly and the spiritual is never categorical. Instead it is mapped according to the constructed "worldly," typically represented by some "carnal" other. The tale of the beggar and the ascetic (told to me by Muni Dulharaji, an aged monk) exemplifies this well:

> A notorious beggar in the neighbourhood was not successful at obtaining food. Every day he would beg at many homes, but the women would rarely give him anything. One day he saw a monk approaching one of the homes he had often gone to and, to his amazement, he saw the woman of the house was eager to feed him. He was equally amazed to observe the monk's bizarre behaviour: he refused most of the things she offered, and of those he accepted, he took only the tiniest quantity. The beggar had an idea. He thought that if he disguised himself as a monk, he too would be fortunate. So the next day he went to the same house dressed in white robes. At first the woman was happy to offer him food, but she soon realized he was an impostor when he eagerly accepted all she offered. His greedy, gluttonous behaviour betrayed him.

Laywomen and men (the beggar being the extreme example) are governed by the transient desires of their bodies; renouncers are governed by their will (expressed in body discipline). The girls at the PSS learn to see evidence of their own spirituality in even the smallest of differences between themselves and householders. As if to encourage this process, the worldly ties are relinquished in fixed stages so that the first-year girls are inescapably more like householders than are the second-year girls, and so on.

Upasika is the name given a girl during her first year. The name means "worshipper," and it denotes the girl's piety and devotion, but doesn't yet earmark her for the ascetic life. Only after the introductory year, if she still wants to pursue the ascetic path, and if her superiors consider her eligible, will she move on to the next stage, to that of a *mumukshu*, "one who is desirous of emancipation." She will likely remain as a *mumukshu* for several years before taking initiation into the order. The *upasikas* have the greatest number of possessions (e.g., five saris) and are given the greater lenience (e.g., they can wash their hair up to twice a week). With each year, the girls try to make do with increasingly less, and thereby come to see the distance between themselves and householders as great. They learn to see "worldliness" in terms of possessions, and in the pandering to the demands of the body (for clothes, food, beauty, comfort, relaxation etc.).

In dress, the *upasikas* are only barely distinguishable from the householders who accompany them in and around the school. However much they dislike it, they resemble the worldly householders more than any other in the ascetic community. In the first year the *upasikas* must wear flower-patterned saris when they are at the PSS. They dislike their "pretty" uniforms, since their prettiness identifies them with the worldly pleasures they are so eager to renounce publicly. When the *upasikas* go on their daily visits to the monastery and to the nun's residence, they wear simple white saris with thick colorful trims, resembling those of the *mumukshus*. But even here, their difference is marked: the saris of the *upasikas* are tucked in on the right and draped counterclockwise, distinguishing them from the *mumukshus*, who secure their saris under their left arm and drape them clockwise over their faces. More conspicuously, the *upasikas* do not wear the *kavatchan* (tunic). To an outsider the differences between the two groups seem minor, but they are important to the girls themselves. Within the school, the *upasika* tries to measure up to the *mumukshu*, and the *mumukshu* evaluates and compares herself with the *samani*—a "sister" who has been initiated into the mendicant order. And all (from *upasika* to *sadhvi*) compare themselves, favorably, with the female householder. Irrespective of how devout the laywoman may be, she serves as a potent symbol of the worldly.

The girls at the PSS self-consciously strive to be "different from householders" in bodily comportment. They try to be more careful in the way they walk, sit, talk, eat, and so forth. They straightforwardly assert that the cultivation of difference is one of their goals, and many times I was asked whether the distinctions were as obvious to me as they appeared to them.

The artwork done by the *upasika* and *mumukshu* sisters at the PSS is illuminating. Paintings done by the girls themselves cover the walls of the reading room at the PSS, and deal with one aspect or another of the ascetic life (e.g., paintings of the founder of the Terapanthi order, Acharya Bhikshu, or of the current leaders, or of the Tirthankaras). One large piece of artwork, part painting, part collage, stands out among the rest. It is a depiction of Jain reality from the point of view of the PSS girls. It presents an upward trajectory of five possible states of human existence—beginning at the lowest, that of the female householder. She is depicted at the bottom corner of the painting, wearing a colorful sari standing over a low cooker. A little higher up on the canvas appears a *mumukshu* sister, dressed in a pink-and-white-striped sari and looking upward at the image of a *samani*. Above the *samani* is a *sadhvi*, represented by an actual picture of Sadhvi Kanak Prabha, the head nun of the order. Finally at the top of the painting is an image of an emancipated Jina sitting cross-legged, eyes closed in deep meditation, and a halo over the head.

From the perspective of the ascetic aspirants, laywomen are unavoidably enmeshed in the chaos and "vomit"[4] of worldly existence. The construction of difference is important in itself: the further the girls progress along the ascetic path, the greater the distance between them and the householders should be.

The Disenchanted Self

Even a short time spent at the PSS makes an enormous difference in the girls' ability to adequately assume the role of a *vairagini*. For example, when speaking of their motives for entering the PSS, the *upasikas* often unself-consciously mix worldly and spiritual aims: they wish to transcend the bondage of worldly life, *and* belong to a group of strong, respected women; they are disenchanted with the world *and* long to wear the white saris of the nuns or tunics of the *mumukshus*; they claim detachment from their bodies *yet* they are repelled by the pain of childbirth. By the time these girls are ready for initiation (*diksha*), all this has changed: their motives are expressed explicitly and solely in terms of disenchantment with the world. They master the skill of projecting an air of perpetual introspection, and they are adept at reinterpreting their past in terms of a narrative of disenchantment. They "learn to remember" and reinterpret their past in accordance with a narrative of ascetic detachment. Motives are rephrased and funneled into disdain for worldly existence. As a publication of the PSS puts it, "The inmates of the

PSS have to pass through a series of experiments aimed at bringing about a radical change in their attitudes towards life" (Bhatnagar 1985:82).

Significantly, their body language accompanies this change. Generally, when the girls enter the PSS, they are bashful and modest—female virtues in social life, but unfitting for a *sadhvi.* Their ascetic body language is anything but fluent. Even though proficient at certain aspects (e.g., covering their mouths when speaking, doing *vandana* or salutation to ascetics), their utterances can be spontaneous and their gestures unprompted. By the time they are ready for *diksha,* however, their body language has changed from self-conscious (body-conscious) to self-assured ("possessed of will"). Their body gestures and modesty no longer suggest sexual restraint, but are now desexualized, and speak the language of nonviolence, control, and restraint. Interestingly, these changes are manifested in ways that may at first appear to be a move toward boldness, rather than toward increased restraint. Their timidity appears to have vanished; they speak more forcefully to men and women, without distinction; they walk with longer strides. Meekness, if it appears, is reserved for superiors within the ascetic community.

The renouncers are, of course, more restrained in their comportment. The difference is in the restraint being manifested: Ascetic restraint denotes nonviolence, which is revealed through a fearless, bold body. Feminine restraint denotes sexual respectability, which is revealed through the concealed body. For the *sadhvi,* her body becomes a neutral vehicle for expression. For the laywoman, the body is implicated in the message conveyed; its association with sexuality seems intrinsic and cannot, therefore, be neutral.

The Lotus

A girl's residence at the PSS comes to an end with an invitation to take *diksha* (initiation into ascetic order). By this time, she has become a masterful *vairagini.* For many weeks before the momentous event, her family will indulge her body for the last time: they will shower her with jewelry and saris and others gifts that she will have no use for. She often receives so many new outfits that she changes her clothes several times a day. Her family will take her to visit pilgrimage sites and even local tourist attractions, because her eyes will never again get the chance to see them. They feed her delicious foods that she will never again taste. At the *diksha* event itself, she is dressed as a bride—decorated, garlanded, and painted—so that her beauty can be celebrated for the last time. Throughout it all, the young woman remains indifferent. Before the enormous crowds that gather for the *diksha* ceremony, she shows no emotion. When her mother sobs at the sight of her long hair being shaved off, she remains unmoved. And when her family grasps at her body as she moves away from them, her stride does not slow down. As a *vairagini*, she is already gone.

The body is lavishly indulged before the *diksha* because it will never revel again. Instead, the desiring body—a symbol of the worldly—now becomes the main focus of the *sadhvi's* discipline and meditation.

Meditation upon the "otherness of the body" constitutes a core practice of ascetic discipline and is believed to inhibit the inflow of karma. These meditations constitute one of the mandatory "12 reflections," which all ascetics perform.[5] The Tattvartha Sutra describes it as:

> Reflecting upon he otherness of the soul from the body and other physical objects focuses on the soul as an eternal intelligent self and the body as an evanescent insentient object. The intrinsic purity of the soul is experienced and the bodily attachment obstructing the spiritual path is destroyed. (Tatia 1994:223–234)

Through discipline and meditation, the *sadhvi* becomes aloof from her body. She dwells in it, but she is not of it. Like the beautiful lotus flower that rests upon the swampy dross, she is within her body and within the world, but she remains unsullied.

For the purpose of her soul, the *sadhvi* pays uncompromising and enduring attention to the body. Every dimension of ascetic life can be viewed as a case study in body discipline. Consider, for example, an average day in the life of a *sadhvi* (based on my experiences of living among the Terapanthi ascetic order[6]). Every aspect reveals the authority of the will over the body:

> It is three in the morning, and deep within the walls of the monastery, the *sadhvis* (nuns) begin to wake. The unruffled thin sheets covering their bodies indicate tranquil sleep, an auspicious sign. Completely motionless sleep is the objective because all movement is potentially dangerous to other living beings. Unkempt sheets would indicate reckless gestures, and thereby, transgression. Spontaneous, unmonitored activity leads to the death of countless lives that surround us at all times. The *sadhvis* rise slowly from lying on their backs and settle into a lotus posture on the cardboard mat upon which they had slept. The *muhpatti* (or mouthshield) remains on. It is worn during the night for the same reasons it is worn during the day: viz., to prevent the harm to subtle living beings in the air, and is removed only when eating. Sitting tranquilly, the *sadhvis* begin their prayers and meditation, awaiting the rise of the sun. In the darkness, where the presence of tiny living beings are concealed, movement is folly.
>
> After more than two hours the first hint of light sneaks in through a window. The countdown is on: 30, 20, 10, 5 more minutes . . . finally there is

enough light to make out the fine lines in the palm of one's hand, and activity can begin. By insisting on sufficient light to read one's palm lines, the ascetics ensure their ability to be mindful of the living beings in their proximity. They begin the practice of *pratilekhna*—(the meticulous inspection of one's clothes [and all items] for tiny life forms). Finally, with the sun high in the sky, there is a quick retreat from the building and they are off for the collection of alms. Consuming food is strictly prohibited when the sun goes down, so the items collected this morning will be the *sadhvis'* first bit of food and drink since before sunset the previous day.

As they move swiftly along the village paths in search of alms, they are mindful of each step. They walk only on sand and cement, for walking on grass would mean killing it; brushing against a bush would mean harming it. Jain ascetics cannot prepare their own food. Plants, water, fire, electricity—all the things necessary for cooking—are considered alive. By the time the ascetic consumes the food, it must be devoid of all life. By ingesting food and water that are no longer alive, the ascetic accrues no karma. Through the generosity of the pious householders, the ascetics remain karmically unaffected by violence inherent in the preparation of food. The householder, who has not renounced the world and is living "in" society, accepts that a certain amount of violence is necessary in order to survive, and is more than happy to provide ascetics with alms. By doing so she earns good karma.

They approach the home of a pious lay person. Her door is open and she beckons them in. They enter and one *sadhvi* examines the food before her. Importantly, she asks the woman for whom has she prepared the large quantity of food. The woman explains that it is made for her family—and certainly not with the expectation of giving it as alms to the ascetics. [Lay Jains, the householders, are not allowed to prepare food explicitly for ascetics and should not know in advance that the ascetic will be coming to beg. Instead the two are to meet by chance. To prepare food explicitly for the ascetics would involve them in violence].

And how was it prepared? asks the *sadhvi*. The woman of the house explains in detail that all the plants and fruits were boiled; that the water used in making porridge was first boiled. The nuns stand away from the small fridge in the center of the room—for to brush against it would cause harm to fire-bodied beings in electricity. The *sadhvi* asks the woman if she washed her hands with "raw" water. No, the woman is emphatic, she only allows boiled (i.e., "dead") water to touch her. Satisfied, the *sadhvis* yield their alms bowls and collect a small quantity of food. Then they are off.

After visiting five or six homes, the *sadhvis* have collected enough food, and they head back to the monastery grounds. The food is distributed among their small group and consumed in its entirety. If any is left over (which doesn't happen often, but has happened), it must be buried. Discarded food would become the source of an orgy of violence—insects would swarm in it, dogs would eat the insects and would fight amongst themselves for it—and the ascetics would be implicated in the violence. After their meals, the ascetics

> depart again (in pairs) for "excretory purposes." Because water is alive, ascetics cannot use flush toilets, so they venture away from their dwellings to find a patch of land that is devoid of vegetation. Again, their refuse must be buried.
>
> After sunset ascetics are prohibited from going outdoors because the night air is filled with dew which, like rain, is alive. If it is absolutely necessary to go outside, they must cover their heads with a cloth so that the falling dew or rain (water-bodied beings) will hit the fabric, and not die as a result of impacting directly against their bodies. Before sunset, they recite a prayer of *pratikraman* in which they repent for the sin of violence that may have occurred during the day, and perform *pratilekhna* again before changing into their night clothes. No food or medicines are consumed after sunset and, since ascetics cannot use electricity, they remain in the dark until a householder turns on a light. By nine o'clock most have carefully lowered themselves onto cardboard mats (thoroughly examined for insects) and fallen fast asleep.

The life of a renouncer permits little spontaneity. The body is no longer free. It does not take a single uncontrolled step. Ascetics explain the necessity of taking vows, or oaths, of restraint in order to frame and bring action under the control of the will. They assert that in the absence of vows, nonviolent action is only slightly effective. Many times I was told: "If a man walked across a path and fortuitously harmed nothing, he would attract more karma than a *muni* (Renouncer) who, when walking, accidentally crushed a tiny insect." Action that is free may or may not cause harm, but body freedom indicates lack of restraint and thought. Action that is regimented operates within a framework of restraint, and under the control of the self.

Renouncers act within a field that has already been mentally mapped. The body obeys the will of the self and is no longer an instigator of thoughts, actions, or desires. For most householders, it is the body that reigns supreme. Unpredictable, demanding, and overbearing, it is oblivious to the murmurs of the self. Mindfulness, so central to all *sannyasi* traditions, is the practice that gives the self total dominion over every gesture, movement, and desire.

Conclusion

Most feminist literature is concerned with the body. This is for good reason, given that women's association with the body, if not universal, is extremely widespread across cultures and historical epochs. Indeed, in her book, *Volatile Bodies*, Elizabeth Grosz argues that "[f]emale sexuality and women's powers of reproduction are the defining (cultural) characteristics of women" (1994:13).

Dualism is the practice of differentiating mental from physical dimensions of life, and assigning them opposing values (most typically, prioritizing the mental over the physical, as in Jain and Western traditions). Characteristically, men are associated with the eternal mind, and women are linked with the transient physical.

There are three common feminist critiques of Western dualism. The first is to locate dualism as an intrinsic part of a patriarchal conceptual scheme that needs to be dismantled. The "mind/body" dichotomy is seen as the source of a system that disadvantages women by aligning them with the denigrated physical. Typically, some form of holism is offered as an alternative. Holism would recognize the embodied nature of all life forms, and the metaphysical meaninglessness of dualism.

Another position is to accept dualism but invert the hierarchy: the "mind" becomes subordinate to the body. The alignment of women with the physical is accepted, but the body (in particular, sexuality and fertility) becomes the center of celebration.

A third response is to accept the basic tenet of the relative inferiority of the body, but to reject women's alignment with it. The physical is subordinate to the mental, but women are no more "embodied" than are men.

Jain views on dualism do not correspond to any of these positions. First, dualism itself is at the core of Jain ontology, and therefore it is too fundamental a belief to be in question. Second, female ascetics of the Terapanthi order (and I suspect more generally) have no hesitation in accepting women as more "physical" (in the sense of being more emotional and more nurturing) than men. Yet none would ever consider valorizing the physical. Jain ascetics are, for the most part, conservative guardians of their tradition. All accept the dualism of self and body as real, and would eschew any notion of holism as foolish. A position advocating the valorization of the body would be dismissed as *mithyadrishti* (delusional).

Operating within this ancient worldview, women accomplish remarkable cultural and personal goals. They boast of being more powerful than the strongest or richest householders on earth. Unlike those with worldly powers, they alone possess unconditional power. They hold absolute power over their bodies—controlling its appetites and pleasures, and serene in the face of illness and death. Fears, like desires, have their origins in the body, and when it is silenced, there is no reason to fear. Unlike most, they have dedicated their lives to transcendent goals, and their bodies have complied; they have not lived for bodily pleasures only to awake, as death approaches, to the meaninglessness of worldly life.

* * *

A *sadhvi*'s limp and aged body sits alone in the open courtyard of the nun's residence, propped up against a pillar by a scarf extending around her face.

Five weeks earlier she had taken the vow of *sallekhana*, the fast until death. *Sallekhana* is the supreme expression of a desire to disengage, putting an end to the body's activities that bind it to this world. This evening the old nun became free of the body that had become a burden to her. Her constant pain made her *sadhana* (spiritual practices) impossible to perform. For many months, she had to rely on other nuns to collect alms, to dress, and even to help her recite lengthy prayers.

The tool that had served her in her spiritual development now threatened to take over her life. Rather than be immersed in spiritual practice, she found herself devoting too much time to its consideration. The "reappearance" of the body's voice threatened to consume the *sadhvi* in worldly concerns. This would be a tragic return to the reign of the hegemonic body—not through the youthful association with sexuality and family life, but now through its demands for relief.

Because the purpose of the *sadhvi*'s life was not tied up with the caring and comforting of others, she was free to sever her ties with the world. Unlike her lay sisters, no earthly responsibilities beckoned her, nor did they take precedence over her own spiritual needs. Her body was not a tool for the well-being of others, it was hers alone and therefore her decision whether to nurture or abandon it.

The *sadhvi*'s body is a thing of the world, and therefore it is of no importance to the ascetics. Soon, it would be removed from the nun's residence and returned to "society" where householders would provide the proper cremation rites.[7] This final handing over of the body is symbolic, making concrete the ascetic's repudiation of the body and of worldly existence.

Determining the end of the relationship with one's body is the most powerful expression of the authority of the soul. It is a rare and auspicious event. For most living beings, it is the body that astonishes the soul with its demise—an unfavorable end because disorientation causes the soul to panic, producing intense emotions that bind deleterious karma to the soul. The *sadhvi*'s body did not die in a state of bewilderment or ignorance; not in the pain of childbirth, in a senseless, capricious accident, nor engrossed in the agonies of the flesh. Instead, she approached the death of her body with awareness and indifference. It was the ideal end to life of restraint and nonviolence, and one that all renouncers aspire to. She ended her life as she lived it: with discipline, with purpose, and with ultimate concern for her soul.

Her hands were not for henna; they were tools of nonviolence. Her feet were not for scurrying after little ones. They were for steady steps along a pious path. Her body was her instrument; it is not her. And when it served her no more, she ended her connection with it forever.

Notes

1. According to Jain thought, the soul may have up to five types of bodies associated with it: namely the gross, protean, conveyance, fiery, and karmic. Some bodies exist to provide supernatural powers (e.g., the protean and conveyance bodies), but they are acquired only in an advanced spiritual state. The fiery body, which everyone possesses, serves the function of digestion. The karmic body is also intrinsic. It carries the karma from one incarnation to the next. Throughout this essay, when I refer to the body, I am referring to the "gross" or physical body, and not to any of these additional "bodies" (see Tatia 1994:55).
2. The Acaranga Sutra is one of the most ancient and authoritative of Jain sacred texts.
3. The situation was different in past generations when many girls were sent off to be nuns to avoid the cost of dowry (see Reynell 1985). Today the vast majority join on their own volition.
4. A common term used by renouncers to describe worldly existence.
5. The other 11 reflections are upon the impurity of the body, impermanence, helplessness, the cycle of birth and death, solitariness, inflow of karma, wearing off of karma, the nature of the cosmos, rarity of enlightenment, and the lucid exposition of Jain doctrine.
6. This section is paraphrased from Vallely 2001b.
7. Unlike Hindu renouncers, Jain renouncers are cremated. Jains do not accept the Hindu idea that the body of a "*siddha*" has achieved a level of purity that makes the purifactory ritual of cremation unnecessary. Instead, they maintain that the physical body, whether of a layperson or renouncer, is always thoroughly a thing of the world. With death, it no longer contains a spiritual essence and is to be disposed of according to social custom. For Jains, death rites are social rituals that have no effect on the state of the soul or on the soul's continued spiritual journey.

Bibliography

Babb, Lawrence.1996. *Absent Lord: Ascetics and Kings in a Jain Ritual Culture.* Berkeley: University of California Press.

Balbir, Nalini. 1994. Women in Jainism. In *Religion and Women.* A. Sharma, ed., pp. 121–138. Albany: State University of New York Press.

Banks, Marcus. 1986. Defining Division: An Historical Overview of Jain Social Organization. *Modern Asian Studies* 20(3):447–460.

———. 1997. Representing the Bodies of Jains. In *Rethinking Visual Anthropology.* M. Banks and H. Murphy, eds., pp. 216–239. New Haven: Yale University Press.

Bhatnagar, R., ed. 1985. *Acharya Tulsi.* Ladnun: Jain Vishva Bharati.

Cort, John. 1991. The Svetambar Murtipujak Jain Mendicant. *Man* (n.s.) 26:651–671.

Clementin-Ojha, Catherine. 1987. Outside the Norms: Women Ascetics in Hindu Society. *Economic and Political Weekly*, April 30:34–36.
Dube, Leela. 1988. On the Construction of Gender: Hindu Girls in Patrilineal India. *Economic and Political Weekly*, April 30:11–19.
Dundas, Paul. 1992. *The Jains*. London and New York: Routledge.
Folkert, Kendall. 1987. Jainism. In *A Handbook of Living Religions*. J. Hinnels, ed., pp. 256–277. Middlesex, England: Penguin Books.
Goldman, Robert. 1991. Foreword. In *Gender and Salvation: Jaina Debates on the Spiritual Liberation of Women*. By Padmanabh Jaini, pp. vii–xxiv. Berkeley: University of California Press.
Goonasekere, S.A. 1986. Renunciation and Monasticism Among the Jains of India. PhD thesis, University of California.
Granoff, Phyllis, ed. 1993. *The Clever Adulteress and Other Stories*. Delhi: Motilal Banarsidass.
Grosz, Elizabeth. 1994. *Volatile Bodies: Toward a Corporeal Feminism*. Indiana: Indiana University Press.
Haug, Frigga. 1987. *Female Sexualisation*. E. Carter, trans. London: Verso.
Holmstrom, Savitri. 1987. Towards a Politics of Renunciation: Jain Women and Asceticism in Rajasthan. MA thesis, University of Edinburgh.
Jaini, Padmanabh. 1991. *Gender and Salvation. Jaina Debates on the Spiritual Liberation of Women*. Berkeley: University of California Press.
Kerns, Virginia, and Judith Brown. 1992. *In Her Prime: New Views about Middle-Age Women*. Urbana: University of Illinois Press.
Reynell, Josephine. 1984. Equality and Inequality. Unpublished paper presented at seminar on Jain Religion and Society, Rajasthan University, Jaipur, August/September.
———. 1985. Honour, Nurture and Festivity: Aspects of Female Religiosity amongst Jain Women in Jaipur. PhD thesis, University of Cambridge.
Shanta, N. 1985. *La Voie Jaina: Histoire, Spiritualité, Vie des ascetes pèlerines de l'Inde*. Paris: OEIL.
Tatia, Nathmal, trans. 1994. *Tattvartha Sutra ("That Which Is")*. By Umasvati. The Sacred Literature Series. San Francisco: Harper Collins Publishers.
Vallely, Anne. 2001a. Ambiguous Symbols: Women and the Ascetic Ideal in Jainism. In *Feminist (Re)Visions: Landscapes, Ethnoscapes and Theoryscapes*. Gail Currie and Celia Rothenberg, eds, pp. 131–144. New York: Lexington Press.
———. 2001b. Moral Landscapes: Ethical Discourses among Orthodox and Diaspora Jains. In *A Reader in the Anthropology of Religion*. Michael Lambek, ed., pp. 555–569. Oxford: Blackwell Press.
———. 2002. *Guardians of the Transcendent: An Ethnography of a Jain Ascetic Community*. Toronto: University of Toronto Press.

Chapter 10

Afterword: Breaking Away . . .

Ann Grodzins Gold

Close engagement with the essays collected here—a set of writings richly conceived, complexly interwoven, and ethnographically substantial—has expanded my knowledge, dislodged false assumptions, and opened new horizons. This work's total impact led me to wonder, as other readers may, whether the concept of renunciation in South Asian religions loses all coherence when we explore its multiple interpretations in such diverse contexts and stupendously messy realities. Where I am able to gather my thoughts is at the indeterminate yet momentous juncture of "breaking away."

In opening their comprehensive introduction to these essays, Hausner and Khandelwal speak of our coedited volume as a work about female religious practitioners who have "broken away from the social roles and structures expected of them" in their respective cultures. Yet to contemplate these case studies collectively is to see clearly that in each one the terms on which women negotiate their breaks and the conditions in which they live them vary enormously. Each author shows us real persons fully enmeshed in the existential and interpersonal complexities of human existence as well as the politics of identity. What does it mean then for a woman to "break away"? Each of our authors answers this question vividly and differently.

Most of the women we meet in these pages are involved in, or reflecting on, processes of disengagement and disentanglement, some ongoing and some more fully achieved. Some breaks may be celebrated ceremoniously and accepted as irrevocably decisive, while others comprise more gradual, shifting realignments. For Jain nuns, whose cosmology Vallely shows to be profoundly dualistic, the two sides of the break may be night and day. But the essays assembled here could lead readers to conclude that for most women renouncers in other South Asian traditions the boundaries between a renouncer's life dedicated to spirituality and a householder's life enmeshed in worldliness and domesticity are almost impossibly fuzzy, blurry, and permeable.

Take Radha Giri in Hausner's lively description, who camps permanently in a public space and comports herself in such a fashion that modest householder women might shun her company, saint or not. Yet by undertaking child-rearing, she participates insouciantly in a paradigmatic female householder's role. Listen to Baiji in Khandelwal's textured portrait, as anthropologist and *sannyasini* companionably chop cauliflower: "You'll write this in your thesis? Sadhus do what sort of work? Householder's work." Together, ethnographer and saint, two women, share a good laugh. Attend to Tara in Hanssen's intimate account flatly stating, "You are not a sadhu if you do not have a spouse," thus contradicting from her Bengali Baul viewpoint cherished preconceptions many hold about practices of religious renunciation in South Asia, or anywhere.

The thoughtful and articulate Baul women in Knight's essay similarly confound notions of what renunciant females are and are not. Among them, Nur Jahan, a Muslim, finds only the poetics of madness can describe her behavior while immersed in a spiritual quest. In Menon's sensitive and restrained ethnography, we see initiated women renouncers who live their religious commitments in and for the public sphere, rather than at a remove from it. They justify this engagement with reference to the Gita's yoga of action (I shall refrain from commenting either on their motivations or on their politics).

Both Shneiderman's and Gutschow's essays, set in the mountainous north, and describing Bonpo and Tibetan Buddhist renouncers respectively, show us women who have been initially dedicated to religious paths by accidents of birth order rather than vocation, and who have spent much of their adult lives contributing labor to their natal households. For past generations of Bonpo renouncers in Mustang, Nepal, as Shneiderman describes them, celibacy was not central to their identity, nor were illegitimate children a blemish on their public character. Chomo Khandru, in Shneiderman's tribute, becomes a traveling businesswoman on her family's behalf, cross-dressing for protection. This worldly role falls upon her—in an ironic twist—because as a renouncer she is not tied down by domestic duties such as cooking and child care. Perhaps only the exhausted saintly selflessness of Baiji in Khandelwal's empathic and frankly admiring portrait, and the exquisite self-denial of the Jain nuns in Vallely's acutely rendered vision, exemplify a female renunciation that might even partially conform to expectations based on common stereotypes of South Asian holy persons such as the loving guru and the isolated ascetic.[1]

This book challenges its readers to understand each female path of renunciation in terms the women themselves have set. The authors have lived with, listened to, recorded, and respected these women's words. While as anthropologists our contributors do not eschew the task of interpretation, each has also ably transmitted women renouncers' own commentary and perspectives on their lives, struggles, and religious commitments. Even as I marvel at the variety of patterns these essays reveal, and speak of fuzzy boundaries, it strikes me powerfully that the religiously dedicated females we encounter in this volume seek to establish their own identities with distinct precision. Blurriness resides in the eye of the outsider-beholder.

The women renouncers we encounter here, as our contributors have carefully portrayed them, crucially know who they are, and why they have traveled their particular roads. Moreover, the society in which they live accepts their breaking away as religiously sanctioned. If social acceptance is more easily granted when breaks are institutionally and ritually marked, it is nonetheless the case for every woman we meet within these pages, that such acceptance is always hard-won—if in notably different ways.

We might design multiple continuums in order to measure various aspects of these renunciant women's behavior and deem them more or less removed from householder women's lives. These might include differences between involvement in and disengagement from public life; maintenance of and nonmaintenance of relationships with natal families; earning and begging for livelihood; enjoying physical comforts and sustaining ascetic practices; subscribing to accepted restrictions on female behavior and transgressing them; and others. I mention such a hypothetical project of assessment as an interesting thought exercise. However, ultimately it could not help us to delineate universal behavioral codes or patterns for they do not exist. Different traditions code different behaviors as more or less appropriate for renouncers: celibacy may be singularly important in one tradition and casually disregarded in another. Hanssen and Vallely show us attitudes toward female bodies that could hardly be more opposed, at least on the surface, whether we think cosmologically or ritually.

Our ability then to group all persons portrayed here as "women renouncers" and to see their lives as significantly different from the lives of women engaged in domestic and worldly pursuits is grounded only partially in specific practices or general modes of existence. Most significant, as I have already indicated, are the ways these women define themselves and the ways society acknowledges the renunciant identities they claim. They have fashioned their lives according to particular regional and cultural versions of what female renunciation might mean. They have named themselves, and are named by others, as set apart. And their agency in doing so is evident even in cases such as Chomo Khandru's where little personal choice was exercised initially. I shall also suggest in this Afterword that there may be women who never break away and who never name themselves renouncers, but who may nonetheless deliberately participate in one or more behaviors characteristic of renunciation. In other words, some householder women position themselves toward the renunciation end of one or more hypothetical continuums without ever directly claiming the label or assuming the full status.

Hausner and Khandelwal highlight in their introduction significant common themes emerging from a collective reading of these essays, while keeping in mind the extraordinary variation each brings into view. They have located our volume's contributions not only within Indological concerns, but more broadly in anthropology, feminist theory, and gender studies. They point to these essays' significant contributions to understanding not only agency but bodies, sexuality, and gendered religiosity both within and beyond the parameters of the South Asian subcontinent. My coeditors' thoroughness frees me to be idiosyncratic—to reflect on our volume's overall impact, not in summary fashion, but rather both retrospectively and prospectively.

Retrospectively, glancing back through my own writings on gender, I consider how South Asian householder women whose words and lives I have sometimes documented do and do not remind me of the renouncer women portrayed here. I hope thus to suggest some ways that these uncommon women's paths may help us think about less radical departures. That is, I suggest that we might—by considering their affinities to housewives, mothers, and daughters—find still broader relevance for these nuns, *yoginis*, saints, and singers. I also tap the images of female religious adepts in oral traditions, both women's and men's, to consider how similar continuities and discontinuities play out in popular imagination. Prospectively, within the several fields this book engages, I highlight a few ways its collective impact might shape future understandings and inquiries.

Retrospective

As someone who has written much about the everyday religious practices of rural Hindu householder women, as well as examining their self-images in expressive traditions, I find that the essays gathered here inspire me to contemplate the ways that householder women may relate their own lives to those of renouncers. In this volume we have met, among others, shrill speech-making *sadhvis* inciting chauvinistic anger in their audiences; nuns whose lives are governed by an excruciating attentiveness to self-restraint and nonviolence; Baiji, the gentle but authoritative *sannyasini* sometimes stressed out by the responsibilities of managing an ashram that seems more like an expansively porous household; an aging unordained Tibetan Buddhist renouncer half-starving in a cave.

Why are such diverse women more like one another than they are like—to foreshadow the example I shall shortly unfold—a householder widow, maintaining her husband's ancestral property, who dresses in the renouncer's color of ochre; wears wooden religious beads; delivers spiritual advice; and spends a considerable part of her day in devotional activities—including ritual worship, hymn singing, and private meditation? Some answers could seem obvious: the latter woman lives with her family, at least most of the time; she has personal property. But others are less clear. For example, if the identity of renouncer allows women to move in public spaces differently from the vast majority who are constrained by roles based on relationships to men (paradigmatically daughter, sister, or wife), then this widow is more like a renouncer. Although her widowhood complicates my argument—and widowhood is a thread to which I shall return in the following section—she fought for and won the freedom to travel for religious reasons when her husband was still alive. If she is involved with private property and finances, then so was Chomo Khandru during her long years as a trader. If she enjoys the company of small children, so does Radha Giri.

Some authors in this volume have argued persuasively that women bound by many culturally sanctioned social constraints find a mode of agency when they choose a path of renunciation. I want simply to add that there may also be women who find it possible to break away, at least partially, without saying so. That is, there may be women who never depart from society or domesticity but whose lives express, in various fashions and to various degrees, a kind of shadow renunciant way of life. These women may deliberately if subtly evoke renouncer models both to seek spiritual fulfillment and to find freedoms beyond given behavioral norms. To exemplify this possibility, I introduce more fully the character I mentioned earlier, to offer our readers one additional example of a real and complicated person from whose words and deeds we have something to learn.

Further Adventures of Shobhag Kanvar

Shobhag Kanvar Chauhan, known to one and all as Bhabhasa, or "honored grandmother," became my landlady in September 1979, when I first moved to the largish Rajasthan village of Ghatiyali. I lived in her sitting room until my departure in the spring of 1981. Although in subsequent fieldwork trips over the last quarter-century I have happened to reside elsewhere, my connection with Shobhag Kanvar has always remained important. In 1987 I elicited from her an oral self-portrait, which I eventually published (Raheja and Gold 1994:164–181). Shobhag Kanvar was a powerful person and she strongly influenced my fieldwork. Not only her life, including her demeanor, activities, and relationships, but her devotional stories—charming imaginative worlds transmitting profound theologies—have in many ways shaped my understandings of Rajasthani women and of Hindu faith (Gold 1995).

Shobhag Kanvar's husband was alive and well, if largely absent, during my first fieldwork period. Well employed in the city of Jaipur, he only returned home for holidays or important life-cycle ritual occasions. By 1987 he had passed away. Except for a strikingly different mode of dress and adornment, it appeared to me that widowhood had little altered Shobhag Kanvar (Raheja and Gold 1994:173). The patterns of her life after widowhood remained the same and included her regular participation in devotion to the Rajasthani hero-God Dev Narayan; her quasi-priestly interventions on behalf of women pilgrims to his nearby shrine; and her attentive upkeep of a home shrine dedicated to Dev Narayan (figure 10.1).

My encounters with Shobhag Kanvar in her widowhood have been intermittent. On each visit I bring gifts that she sometimes disdains (picture frames were not a hit) and sometimes appreciates. She welcomed with pleasure a cotton sweater in a melon color, naming it *bhagva*—a shade conventionally translated ochre, which explicitly characterizes renouncers' dress. Widowed, Shobhag Kanvar had made a clear if idiosyncratic choice to wear nothing but shades of orange. She might, as her previously widowed elder sister, have dressed in green or other dark colors that for the Rajput community—whose married women overwhelmingly wear pink and red—more conventionally serve as widows' weeds. Shobhag Kanvar, however, garbed herself in clothing she identified as ochre. Yet her clothes were in no way exact replicas of renouncers' garments; the cloth was often a flowered print that no ordained renouncer would use. She simply selected fabrics from among the vast assortment of prints available in rural Rajasthan, in which the predominant background was a salmony hue. Thus she suggested through a subtle semiotics of color her affiliation to a renouncer's life even while remaining a householder.

As a devotee of Dev Narayan, Shobhag Kanvar regularly attended weekly Saturday healing sessions and special events throughout the year at the shrine called Puvali ka Devji, on the outskirts of Ghatiyali. The non-Brahmin priest (*bhopa*) of that shrine, Gopiji Gujar, was himself a householder with wife, son, and property. In his later years, however, he went home less and less frequently. Eventually, he took to living at the shrine, and around that time he exchanged his familiar red turban for an ochre one.[2] He never took vows of renunciation nor did he formally abandon his family.

Figure 10.1 Shobhag Kanvar Chauhan beside her home shrine, 2003

Source: Photo by Ann Grodzins Gold.

Yet, when he passed away he was not cremated but buried with pomp in a *samadhi* (renouncer's tomb) just outside the shrine where he had served for so many years.[3]

Shobhag Kanvar herself, even when I first knew her and her husband was still alive, would stay at Puvali ka Devji whenever all-night devotional singing sessions were sponsored by thankful devotees. After her widowhood, she spent more time at the shrine on ordinary days as well as festivals.

In this fashion, as with her orangish clothing, she displayed increasing inclinations for an ascetic life expressive of her deep devotion. While the priest Gopiji was still living, in 1993, an incident took place that shows how Shobhag Kanvar's identity as a renouncer-like woman permitted her to justify behaviors and attitudes that few others would dare.

The shrine of Puvali ka Devji was off the road a bit, surrounded by agricultural land, and about a 20-minute walk from the village center. One night thieves attacked. Shobhag Kanvar, a heavy but vigorous woman by then in her mid-sixties, ran barefooted through the fields from shrine to village to sound the alarm in the middle of the night. The thieves had beaten the priest and stolen his gold earrings. They were miraculously unable to break into Dev Narayan's locked treasury of valuables, and it was left untouched—perhaps because they were forced to flee after Shobhag Kanvar roused the village.

Irreverent village wags composed hilarious doggerel on Bhabha's (Grandma's) midnight sprint; some chattering voices cast aspersions on the reasons for her presence at the shrine that night. But Shobhag Kanvar herself told the story with total aplomb, exercising her considerable narrative gifts to give it high drama and to make herself the perfect heroine—quick thinking in her escape; wounding her unprotected feet on rocks and thorns in an act of impeccable courage and unstinting devotion; thus saving the shrine from worse depredations. She was stonily impervious to whatever gossip fluttered about the incident, and ultimately it died away. Shobhag Kanvar, like all the women in this book, is in many ways an extraordinary personality. I would not want to assert that it was merely her ochre clothes and wooden beads and claims of mystical powers that kept her dignity intact. However, I would say that these attributes, which I am convinced are sincere if also strategic, were certainly components of her ability to remain unperturbed by ripples of public opinion.

Every time I visit Shobhag Kanvar's house, improvements are underway. In 2003 it was a septic tank and flush latrine; earlier she had built a stately arched entranceway to her compound, affording a cool and comfortable place to receive visitors; she had planted a wonderful little orchard of fruit trees—papaya, guava, and lemon; she had constructed comfortable, separate homes for her two sons and their growing families. Her impressive courtyard shrine to Dev Narayan, a fixed feature, was frequently improved and repainted. Shobhag Kanvar had not the slightest sense that her penchant for home improvements might clash in any manner with her devotional practices and her ochre clothing. She presided over multiple worlds with royal nonchalance. I rarely saw her without a grandchild, or in 2003 a great-grandchild, kept affectionately close beside her.

I concluded about this extraordinary woman in 1994, "As her stories teach, a woman who defines patterns that differ from those set by authoritative males may please God and better her own existence too" (Raheja and Gold 1994:178). But, you may wonder, had she broken away? The only possible answer is "yes and no." Shobhag Kanvar deliberately adopted some attributes of renunciation and artfully blended them with domesticity. If she has occasionally aroused gossip (and the episode of the shrine robbery was hardly the first time), it has rolled over her without causing long-term damage. I think this is largely because of her unflagging dedication to Dev Narayan coupled with a crucially strong sense of her own virtue. It is difficult for me to imagine Shobhag Kanvar as separated by hugely significant gulfs from many of the other adventurous, self-possessed, and self-willed women in this volume who have in common, if nothing else, questing souls and strong spirits.

Tales of Women Who Break Away

Conservative rural society is naturally suspicious of independent women, as one of Shobhag Kanvar's devotional stories reveals. The story of the *jungli rani* (queen from the forest), told on the day of Sun Worship, portrays how easily a devoted female who gains divine favors can be mistaken by those around her for a dangerous, threatening, or evil figure who abuses power for selfish reasons (Raheja and Gold 1994:149–163). The *jungli rani* knows who she is and deflects every accusation by repeating her life story straightforwardly.[4] I reported in an earlier essay about this tale that the misunderstood woman is eventually accepted on her own terms. She rejoins society, and lives happily ever after with her husband the king, and her baby. Well, that is the way Shobhag Kanvar tells it.

I had concluded that this women's worship story effectively argues against society's false suspicions of an isolated female's single-minded religiosity. Several years later I learned, not without shock and chagrin, that the same story may resolve itself quite differently. Variants of this tale exist as far away as Kangra in the north. There, strikingly, as Kirin Narayan recorded it, the *jangli rani*, pious though she may be, dies at the end—an outcome Narayan found disconcertingly somber (Narayan and Sood 1997:41–45). While in Shobhag Kanvar's version the queen's problematic mother conveniently and magically turns into a gold statue, in the Kangra version, the daughter orders her killed, and there is a subsequent transformation to gold. It amounts to the same thing, really. Yet, Shobhag Kanvar's telling not only vindicates its heroine but lets her survive and flourish. Moreover, the very thing that saves the *jangli rani* in Shobhag Kanvar's version—the telling of her own story—is what causes her death in the Kangra variant.

May I attribute these different outcomes to a kind of vernacular feminist stance on Shobhag Kanvar's part? I would like to think so. Although the word itself would be foreign to her, she frequently expressed to me both in everyday conversation and in commentary on her own stories a strong sense of women's capabilities and of injustices perpetrated on them. In a footnote to their introduction, Hausner and Khandelwal, following Uma Narayan, distinguish between a critique of gender injustice and a critique of the system that perpetuates gender injustice. Narayan holds that the former would be inadequate for a feminist position. I find it difficult, however, to measure the potency of Shobhag Kanvar's narrative arts according to such a distinction. If she subtly and consistently alters story-telling traditions to portray women in a stronger and more positive light, I would argue that this does constitute a critique of systemic devaluations of women even if it is not articulated in so many words.[5]

Male-authored oral traditions in Rajasthan and elsewhere are quick to imagine how all women might slip readily into roles of untamed destructive selfishness, lured by gurus who are themselves malicious females. Consider the reckless behavior of the seven lady magicians (*jadugari*) in the Nath folk epic of King Gopi Chand, whose leader is a *yogini* named Behri. All seven have day jobs, but they use their occult powers, gained through ascetic practices, to attack male renouncers, enchant, imprison, and abuse them physically. The text makes it very clear how this upside-down female behavior spreads, infectiously, from guru to disciples. There are seven lady magicians, but each is a guru, and each guru has 700 disciples. When they invite their disciples to join them for a kind of grand tournament with male yogis, here is how the bard describes householder women's responses:

> They were dying from anticipation. If one slut was grinding flour, she left the flour in the mill; and if one was rolling bread, she left the flour in the dough-dish, she left the dough: "Later I will roll it." And if one was nursing a boy or girl, she tied him in the cradle: "We will be late for the show, and miss the fun, so let's go now and nurse later." The sluts, all the women of the city, acted this way. (Gold 1992:249)

What we have here is a male view of women breaking away and it is not a pretty one. As Chitgopekar writes of *yoginis* in Sanskritic traditions, they present "a moral risk to patriarchal society" (2002:84). This is exactly why we need so badly to learn the meanings of a female life lived on terms other than domestic from women themselves; hence why the essays in this volume are so precious.

The large majority of women's songs that I recorded in Rajasthan are thoroughly grounded in householder relationships and cravings; they are all about love, desire, jealousy, hurt feelings, jewelry, fine clothes, food, and the yearnings and sorrows all of these provoke. Only a rare few imagine breaking away. I close this retrospective segment with lines from an unusual devotional song addressed to Dev Narayan, Shobhag Kanvar's chosen deity. Joseph Miller recorded Gujar women on pilgrimage performing this song at a temple in 1980; 23 years later I taped a different and younger group of women, neighbors gathered for the pleasure of singing in Bali Gujar's house, performing the same song. The words had not changed.

> Out of anger with my father-in-law, I went out, Lord
> I will be I will be a wild blue cow [*nilgay*, a kind of antelope, the only large wild animal still common in this region]
> I will wander and graze on the forest grass
> I will drink water from the ocean-tank
> I will see at his temple the Brave Devji.[6]

Each line repeats twice, and the entire series repeats multiple times with different relatives' names in the first line: husband's brother, husband's sister, and so forth. In these simple verses women melodically imagine themselves breaking away for religious reasons. The freedom they find from discord in their marital homes is not only the freedom to worship as they choose but to live a totally extra-domestic life. I can speculate that this might be analogous to a renouncer's life. In any case, women's expressive traditions occasionally reveal deep wishes for freedom, however romanticized their portrayal of what it might comprise.

Prospective

The contributions to this volume expand our horizons enormously, stimulating new ways of thinking—not only about women's roles in South Asian societies and religious traditions, but about gender, bodies, religion, and renunciation writ large. In successfully navigating the confluence of these multiple streams of thought, our book addresses the consequences of gendered embodiment in the everyday behaviors and hopes of actual women. These women's lives are enacted and conceived in diverse but contiguous religious and cultural worlds.

Two promising methodological angles are worth emphasizing here as significant. The first has to do with the ways that ethnographies of renunciation open windows in a fashion that religio-historical approaches, however valuable and insightful, cannot.[7] We could call these windows onto "real life" aspects of women's renunciation, and Hausner and Khandelwal have already directed readers' attention to the significance of this approach. Linked to this is our volume's focus on ordinary renouncers—women who have not risen to transnational guru fame. The latter factor inevitably alters the terms on which it is possible to encounter these persons. Accounts of their lives tend to transform the everyday to hagiography or mystery, and the particular to the conventional.

A second striking methodological contribution is what I might call, echoing Fred Eggan's classic essay of the mid-1950s, "controlled comparison" (1954). By this Eggan meant comparison of different tribes' social organizations within a regionally circumscribed area; more particularly these were tribes related to one another as, for example, Zuni and Hopi—both being Pueblo. As I read the essays in this volume, I find it astonishingly interesting and productive of new insights to see how a single practice, that of women's renunciation, viewed in a single geographic region, northern South Asia, may play out within a broad spectrum of religious traditions, cultures, and socioeconomic contexts. We encounter fascinating variations cosmological, theological, and sociological. If common elements emerge, they are all the more striking.

A collection so rich in content stimulates not only new reflections on what we thought we already understood, but new, synthetic, and breakthrough directions for further explorations. I conclude this afterword by sketching some fruitful areas beyond the present volume's scope but within reach of its generative arguments and strategies. Three of these are comprehended in a single theme that emerges from my retrospective comments—a theme I shall awkwardly describe as continuities and discontinuities between women householders and women renouncers. Under this heading I distinguish three topical foci: (1) gazing in both directions; (2) widows as renouncers; (3) bodies, modesty, limits, and freedoms. In concluding I shall touch on one other pervasive motif or subtext that stands out most clearly in its absence.

Women Householders and Women Renouncers . . . in Mutual Regard

Most of our essays focus on renouncer women's own views of their lives, although most also evoke at times a gaze from outside. Menon's is the sole essay written almost totally from a spectator viewpoint; in her essay, with one minor exception, the politically active *sadhvis*' everyday lives and personal histories remain opaque. From Menon we learn a great deal, however, about how young women drawn into a political movement may regard with adulation (or less frequently skepticism) those activist women renouncers who exhort them. The *sadhvis*' power to inspire awe and to arouse violent emotions is apparent and suggests how further inquiry into such responses would be of value.

Other essays show us multiple ways women renouncers are engaged in conversations and transactions with women householders. Knight, for example, is concerned with Baul women's views of themselves, but remains acutely sensitive to their consciousness of public opinion. In her portraits of Baul women whose community deems marriage to be a condition for valid renunciant behavior, but who themselves have no stable relationships with male partners, we see how single women are particularly vulnerable. For their livelihoods as beggars and as performers Baul women are dependent on householder responses, as Hanssen also shows.

Jain nuns beg each day as well, but the householder women who feed them consider it a privilege, and must pass the nuns' stern tests before their offerings are received, a choreographed reversal of most begging situations. In the Himalayan region where Gutschow worked, Buddhist nuns are most often supported not by the community but by their natal families, for whom they continue to provide important agricultural labor. Yeshe in Gutschow's essay alienates her relatives with outbursts of her "sharp tongue" and suffers the economic consequences. Readers may be curious as I am just what those angry relatives really think about Yeshe, and why she did not hold her tongue, given her dependency on their good graces.

In general, then, this volume suggests to me that it would be fruitful to know more about how women who remain with their feet well planted in the householder's universe of values and behaviors think not only about specific women who have broken away but about the whole project of leaving the world. Besides the cases of individuals like Shobhag Kanvar who subtly evoke renouncer models to justify their own idiosyncratic behaviors, there are other more common and pervasive resonances. Vatuk, for example, noted decades ago that aging itself for ordinary women often involves a gradual withdrawal and relinquishment, a transition to classical Hinduism's third life stage of "forest dweller"—enacted without necessarily going to the forest (Vatuk 1980, 1990). If handing over to daughters-in-law the keys to storerooms is one symbolic form of renunciation for householder women, going on pilgrimage is another (Gold 1988). If householder women, enmeshed in relationships and answerable to myriad demands for love and work, fantasize the freedom of a forest life, how do they actually feel about women who have achieved that freedom?

. . . As Widows

Our volume has not paid much attention to widows, perhaps because most of our essays are in one way or another about how women choose and forge paths of renunciation. Widowhood seems to imply accident, a lack of choice, fatality transformed directly into fate. Yet there are many strong continuities, both conceptual and physical, between female renouncers and widows, and at times the two categories merge. For example, taking vows of renunciation and residing in an ashram or religious community is one common path for widows both young and old.

In recent years, a spate of publications has enhanced historical, sociological, and emotional understandings of South Asian widowhood, and made widows' own complex feelings far more accessible. Collected documents on widowhood have revealed a range of experiences and commentaries through the centuries.[8] A 1924 article by Gandhi, included in a recent compelling anthology that gathers both documents and fictions, gives us his view of widows as "renunciation personified." Gandhi writes (oblivious to any irony that may strike readers years later): "Self-control has been carried by Hinduism to the greatest height and, in a widow's life, it reaches perfection. . . . A great many widows do not even look on their suffering as suffering. Renunciation has become second nature to them, and to renounce it would be painful to them. They find happiness in their self-denial" (2001:122–123). Self-suffering, Gandhi's famous reformulation of renouncers' ascetic practices or *tapasya*, is to him epitomized by the classically construed Hindu widow. In spite of his manifold and revolutionary commitments to social justice and reform, Gandhi is clearly reluctant here to propose altering the conditions of widows' lives. His most radical suggestion is that men who are widowed should also be prevented from marrying again!

A counterexample to this idealization of widowhood as a culminating pinnacle of renunciant virtues and passionately embraced self-suffering is found in subaltern historian Guha's famous essay, "Chandra's Death" (1997). Here too, renunciation is associated with widows, but in a wholly different fashion. Guha uses archival records to trace a court case involving the botched abortion death of a pregnant widow. Guha describes her as confronted with two stark alternatives posed by male authorities: "abortion or *bhek*" (51). *Bhek* here refers to joining the community of Vaishnava renouncers which Guha bleakly characterizes as "limbo for all the dead souls of Hindu society" (53). He tells us, moreover, that a large part of this group was "women excommunicated for their deviation from the approved norms of sexual conduct," and that the largest group of female outcastes was

"Hindu widows ostracized for defying the controls exercised on their sexuality by the local patriarchies" (53). The living conditions of such women, at least as Guha reports them, drawing from archival documents in Bengali from the mid-nineteenth century, are appalling.

Guha is acutely aware that in his essay there are no women's voices excepting that of the person who administered medicine intended to induce abortion. She hoped to provide Chandra an alternative to *bhek* by terminating her pregnancy, and her statement claims only that she "did not realize that it would kill her" (60). Was the life of Vaishnava women renouncers as grimly depraved and exploitative as it appears in the descriptions of "a sympathetic and acute observer" on whose testimony Guha largely relies (54)? We cannot know, but can imagine that there is more to be learned about these renouncers as well as the institutions that sheltered them. Guha's essay reads bleakly to demonstrate widows' total lack of voice and agency, to characterize renunciation as no choice at all—thus in significant ways quite opposed to Gandhi's notion of widowhood as self-suffering. Yet both male observers premise their different understandings on a shared assumption of limited alternatives, which the contents of our volume suggest should be thrown into question.

Sarah Lamb's recent ethnographic work on aging in present-day rural Bengal suggests that the "cutting of maya" or disconnection from love is a major theme—one that becomes particularly acute for widows but that is pervasive among the elderly. This too is a kind of breaking away. Yet, Lamb stresses, "techniques of renunciation were not practiced in old age to oppose or negate life in the world" (2000:141). Some aged widows, as Lamb observed them, felt free to demand comforts, to speak their minds, and generally to ignore rules of decorum that constrain younger women. In these outspoken practices they remind me of some of the women described in this volume. In sum, the complex experiences of widowhood vis-à-vis multiple routes of renunciation, demand more attention.

. . . Their Bodies and Modesty

The place of gendered bodies in religious studies and the anthropology of religion has been well explored in the last quarter century as my coeditors discuss in the introduction.[9] Attitudes toward female bodies and their reproductive functions, as well as toward bodies in general are all of interest when we consider the differences between householder and renouncer women. Here I want to highlight one South Asia-specific element that calls for closer examination. This has to do with postures of modesty. I find it particularly fascinating because of the range of attitudes displayed in our contributors' field sites.

In many South Asian religious and cultural contexts, including Hindu, Muslim, and Jain, a nuanced assortment of meanings gather around concepts of modesty and shame, explicitly linking women's comportment with men's honor.[10] All these evidently serve to limit women's actions in large and small ways, as women themselves are fully aware. All have something to do with the way women comport themselves, and the ways they are trained from girlhood to discipline their own bodies and minds. As Gutschow's and Shneiderman's essays make clear, the Tibetan cultural universe diverges significantly from these others in this regard, and this carries over to issues of celibacy and of women's extra-domestic roles. I speculate this may have something to do with its gender hierarchy not being infused with that potent combination of shame and honor.[11]

The sixteenth century Rajasthani princess-turned-poet-saint, Mira Bai, never said she was a renouncer. Rather she said she was married to Krishna and could not, therefore, accept a human husband. Hawley and Juergensmeyer (1988) cite the early seventeenth century "Garland of Devotees" as containing the "oldest extant hagiographical statement" about Mira. In their translation of the author Nabhadas's verse, are these lines:

> Mira unraveled the fetters of family; / she sundered the chains of shame to sing of her mountain-lifting Lover and Lord.
>
> . . .
>
> She cringed before none: she beat love's drum. /Mira unraveled the fetters of family; / she sundered the chains of shame to sing / of her mountain-lifting Lover and Lord.

We could hardly get a more explicit and dramatic image of breaking away—used here to idealize a female saint who refused to accept a woman's chartered destiny. She rebelled not only against domestic duties but against modesty itself: she "sundered the chains of shame."

But how is Mira's behavior viewed by householders? Both Harlan (1995) and Mukta (1994), in separate studies of present-day attitudes toward Mira Bai, find that whatever adulation she inspires all these centuries later, her behaviors are viewed with nervous distaste, at least within her own Rajput community. She is the opposite of a role model, and no one names their daughters after her. Mukta, however, finds farmers and other poor people identify more freely with Mira's struggle. She is, of course, worshipped in temples.

Rajput women, interviewed by Harlan, admired Mira, but always qualified their admiration, setting her apart from themselves: Mira "abandoned everything," one tells Harlan, but then adds, "A Rajput woman wouldn't dare go out; everyone can't be like Mira" (1995:208). Another confesses, "Maybe because I can't step out, I admire her so. She was not a *pativrata* [husband-devotee] though . . ." (208). Thus they carefully inscribe boundaries between their own world views and Mira's.

Male Rajputs may regard Mira with even less tolerance. Mira's legend includes repeated attempts by her in-laws to murder her, including sending the famous "poison cup" that Lord Krishna transformed to nectar. A man from Chittor, the former royal seat that Mira fled, said to Parita Mukta, a scholar researching Mira in the 1990s, "What is so great about Mira that you want to write about her? Why are you traveling around in the way that you are doing to find out about her? . . . Mira did not keep to the decorum of a Rajput princess. Therefore, whoever sent her the poison was not at fault. He did the right thing. Even today, if my wife did something to overturn the prestige . . . of my family . . ." The speaker's wife who is present at the interview, Mukta notes, gives "a start" at this juncture (1994:180–181). Such unabashed venom is startling to readers as well, reminding us that obstacles to women's breaking away may feature coercive tactics beyond the psychological.

Mira's unrestrained ecstatic love of Krishna seems to have little in common with the austerity of Jain nuns' well-modulated lives. We nonetheless observe convergence when Vallely notes that the self-realized, initiated nuns no longer need to use modest body language because they have totally separated themselves from everything worldly. Radha Giri in Hausner's portrait similarly has no use for poses of female modesty.

Not all the women renouncers portrayed in this volume have arrived at a space, physical or emotional, where they may joyfully sunder the chains of shame. The Baul world view with its Tantric underpinnings may celebrate female bodies and value menstrual blood in radical ritual action. Yet Baul women, as Knight shows, feel society's pressures strongly. Nor does the absence of restrictions based on shame necessarily result in greater empowerment for females. In the Tibetan cultural context, where shame and modesty are clearly less salient factors, other kinds of devaluation limit women renouncers' lives or consign them to deficient livelihoods. I imagine an intensive comparative focus on ideas about shame and modesty in women renouncers' and women householders' lives across traditions might capture some revealing and patterned variations.

Beyond the Limits

As social scientists—even if irredeemably on the soft or qualitative side—anthropologists are inclined to look at women renouncers in terms of how they break away from social constraints and prescribed domestic duties; from patriarchal structures and all that these entail; from the ways women's bodies are identified with sexual and reproductive functions. We get closer in this volume than in any other studies of which I am aware to female renouncers' everyday lives. In the process I have occasionally wondered whether we had lost track of what all these variants of renunciation signify in the religious universes that inspire them. Liberation, enlightenment, transcendence, self-realization, supreme truth are terms and phrases we encounter, but only rarely, in this book. Our contributors have not lingered much over their implications, stepping lightly, gingerly if respectfully, around their meanings.

Even in the intimate shared space of long-term ethnographic fieldwork, such matters are difficult to communicate; or, as a truism of esoteric traditions holds, can only be passed from master to disciple in direct transmission or training. Our authors do not present themselves as disciples, yet in various ways, at various times, some may have assumed that role and perhaps taken it seriously. Hanssen learns to sing the Bauls' mystical songs; Khandelwal participates in *seva* at Baiji's ashram. Subtle as these discipleships may be, all of our authors, excepting Menon for good reasons, reveal themselves to have been personally affected by their close and deep encounters with South Asian women renouncers. Yet all have needed to maintain a level of analytic separation that creates in their accounts a kind of absence or black box around deeper religious meanings. Here my prospective project is not to incorporate spiritual quests and truths into ethnographic knowledge, for I understand the valid reasons this would not be possible. Nonetheless, we deliberately placed Vallely's chapter at the end to signify not closure but jumping off into a wider unknown. For in describing a nun's death in a meditative state, the victory over flesh and mortality itself, she returns us to another sense of breaking away which informs all South Asian women renouncers' lives.

Notes

1. I draw on my everyday experience of popular imagination in the United States to call up two disparate figures that became prominent in the 1960s: the beatific guru, perhaps modeled on ever-smiling Maharishi Mahesh Yogi; and the isolated ascetic, a cartoon yogi on the mountain top. See also Narayan 1993.
2. In this, he had a kind of local precedent in members of the Nath community, who by ancestry are understood as "householder yogis" and who wear ochre turbans. Among the Naths of the village Ghatiyali was an important devotee at the shrine who was the lead singer of Gopiji's "bhajan party." Nath women do not, however, regularly dress in orange to mark their identities, although they do sometimes develop magical expertise (Gold 1992).

3. See Khandelwal (2004:1) for a discussion on why renouncers are buried and not cremated.
4. See Ramanujan who writes, "I'd suggest that agency in these women's tales is connected with their being able to tell their own story and its being heard" (1999:426).
5. For rural women's critiques of gender hierarchy, see Gold 1995; Gold 2002. See also pioneering Indian "feminist" Madhu Kishwar's bracing statement, "A Horror of 'Isms': Why I do not Call Myself a Feminist" (1999:268–290).
6. I originally translated this song, as recorded by Joseph Miller, for an Indiakit distributed by South Asia Outreach at the University of Chicago in 1981. I was moved to hear it again in 2003, and to learn that it was still part of young housewives' repertoire.
7. There are few precedents for such ethnographies and among those the most recent and ethnographically rich are the larger works of our own contributors: Hausner 2007; Knight 2005; Gutschow 2004; Khandelwal 2004; Hanssen 2002; Vallely 2002.
8. Besides Chakravarti and Gill's comprehensive compendium (2001), see also Chen 1998.
9. See also Castelli 2001; Coakley 2000; Law 1995.
10. Some definitions from McGregor's (1997) *Hindi-English Dictionary: laj* (shame, sense of decency or modesty, bashfulness, good name); *sharm* (bashfulness, embarrassment, modesty); *sharamindagi* (sense of embarrassment, feelings of modesty).
11. See also Kathryn March's wonderful book on Himalayan Tamang women (2002).

Bibliography

Castelli, Elizabeth A., ed. 2001. *Women, Gender, Religion: A Reader*. New York: Palgrave.

Chakravarti, Uma, and Preeti Gill, eds. 2001. *Shadow Lives: Writings on Widowhood*. New Delhi: Kali for Women.

Chen, Martha Alter, ed. 1998. *Widows in India: Social Neglect and Public Action*. New Delhi: Sage.

Chitgopekar, Nilima. 2002. The Unfettered Yoginis. In *Invoking Goddesses: Gender Politics in Indian Religion*. Nilima Chitgopekar, ed., pp. 82–111. New Delhi: Shakti Books.

Coakley, Sarah, ed. 2000. *Religion and the Body*. Cambridge: Cambridge University Press.

Eggan, Fred. 1954. Social Anthropology and the Method of Controlled Comparison. *American Anthropologist* (n.s.) 56:743–763.

Gandhi, M.K. 2001[1924]. Renunciation Personified. In *Shadow Lives: Writings on Widowhood*. Uma Chakravarti and Preeti Gill, eds., pp. 122–125. New Delhi: Kali for Women.

Gold, Ann Grodzins. 1988. *Fruitful Journeys: The Ways of Rajasthani Pilgrims*. Berkeley: University of California Press.

———. 1992. *A Carnival of Parting: The Tales of King Bharthari and King Gopi Chand as Sung and Told by Madhu Natisar Nath of Ghatiyali, Rajasthan*. Berkeley: University of California Press.

———. 1995. Mother Ten's Stories. In *Religions of India in Practice*. Donald S. Lopez, Jr., ed., pp. 434–448. Princeton: Princeton University Press.

———. 2002. Counterpoint Authority in Women's Ritual Expressions: A View from the Village. In *Jewels of Authority: Women and Textual Tradition in Hindu India. Laurie L. Patton*, ed., pp. 177–201. New York: Oxford University Press.

Guha, Ranajit. 1997[1987]. Chandra's Death. In *A Subaltern Studies Reader: 1986–1995*. Ranajit Guha, ed., pp. 34–62. Minneapolis: University of Minnesota Press.

Gutschow, Kim. 2004. *Being a Buddhist Nun: The Struggle for Enlightenment in the Himalayas*. Cambridge, MA: Harvard University Press.

Hanssen, Kristin. 2002. Seeds in Motion: Thoughts, Feelings and the Significance of Social Ties as Invoked by a Family of Vaishnava Mendicant Renouncers in Bengal. PhD dissertation, Department of Anthropology, University of Oslo.

Harlan, Lindsey. 1995. Abandoning Shame: Mira and the Margins of Marriage. In *From the Margins of Hindu Marriage: Essays on Gender, Religion, and Culture*. Lindsey Harlan and Paul B. Courtright, eds., pp. 204–227. New York: Oxford University Press.

Hausner, Sondra L. 2007. *Wandering in Place: The Social World of Hindu Renunciation*. Bloomington: Indiana University Press (in press).

Hawley, John S., and Mark Juergensmeyer. 1988. *Songs of the Saints*. New York: Oxford University Press.

Khandelwal, Meena. 2004. *Women in Ochre Robes: Gendering Hindu Renunciation*. Albany: State University of New York Press.

Kishwar, Madhu. 1999. *Off the Beaten Track; Rethinking Gender Justice for Indian Women*. Delhi: Oxford University Press.

Knight, Lisa. 2005. Negotiated Identities, Engendered Lives: Baul Women in West Bengal and Bangladesh. PhD dissertation, Department of Anthropology, Syracuse University.

Lamb, Sarah. 2000. *White Saris and Sweet Mangoes: Aging, Gender, and Body in North India*. Berkeley: University of California Press.

Law, Jane Marie, ed. 1995. Religious Reflections on the Human Body. Bloomington: Indiana University Press.

March, Kathryn S. 2002. *"If Each Comes Halfway": Meeting Tamang Women in Nepal*. Ithaca: Cornell University Press.

McGregor, R.S. 1997. *The Oxford Hindi-English Dictionary*. Delhi: Oxford University Press.

Mukta, Parita. 1994. *Upholding the Common Life: The Community of Mirabai*. New Delhi: Oxford University Press.

Narayan, Kirin. 1993. Refractions of the Field at Home: American Representations of Hindu Holy Men in the 19th and 20th Centuries. *Cultural Anthropology* 8(4):476–509.

Narayan, Kirin, and Urmila Devi Sood. 1997. *Mondays on the Dark Night of the Moon*. New York: Oxford University Press.

Raheja, Gloria G., and Ann Grodzins Gold. 1994. *Listen to the Heron's Words: Reimagining Gender and Kinship in North India*. Berkeley: University of California Press.

Ramanujan, A.K. 1999. A Flowering Tree: A Woman's Tale. In *The Collected Essays of A.K. Ramanujan*. Vinay Dharwadker, ed., pp. 412–428. New Delhi: Oxford University Press.

Vallely, Anne. 1999. *Guardians of the Transcendent: An Ethnography of a Jain Ascetic Community*. Toronto: University of Toronto Press.

Vatuk, Sylvia. 1980. Withdrawal and Disengagement as a Cultural Response to Aging in India. In *Aging in Culture and Society: Comparative Viewpoints and Strategies*. C. Fry, ed., pp. 126–148. New York: Praeger.

———. 1990. "To Be a Burden on Others": Dependency Anxiety among the Elderly in India." In *Divine Passions: The Social Construction of Emotion in India*. O.M. Lynch, ed., pp. 64–88. Berkeley: University of California Press.

GLOSSARY

The chapters in this volume contain words from several languages: Bengali, Hindi, Nepali, Sanskrit, and Tibetan. Many Sanskrit and Sanskrit-related words are used in Bengali, Hindi and Nepali, so there is much overlap in the vocabulary related to renunciation in these Sanskrit-related languages. Acknowledging this overlap, we include Hindi, Bengali, and Nepali terms in a single list for Sanskrit-related languages. Because Tibetan and Sanskrit belong to different language families, we list Tibetan terms in a separate list. Where alternative Sanskritized spellings appear in different chapters, we indicate both variants (*samsar / a; jnan / a; tapas / ya; vairag / ya*). Three Arabic-derived words (*khilaphat, majzub, mazar*) are included in the first list, as they are used by Bengali-speaking Bauls.

Sanskrit-Related Languages of Hindi, Bengali, and Nepali

acharya	Sanskrit scholar; spiritual leader of renouncer group
ahamkar	egotism
ajiv	insentient matter
akhand Bharat	united India
aryika	noble woman; nun of the Jain Digambar sect
asan	seat or posture for meditation
ashram	hermitage in Hinduism
atman	soul in Hinduism
avadhut	ascetic believed to be above all rule
baba	an old man; an ascetic
bairagi(f. bairagini)	Vaishnava renouncer
bansuri	flute
bhakti	devotion in Hinduism
bhandara	feast
Bharat	India

bhek	initiation into renunciation, usually Vaishnava or Baul
bhek-dhari	person who has taken *bhek* or wears clothes signifying *bhek*
bhog	enjoyment; suffering; experiencing the result of good or bad deeds
brahman	ultimate reality; impersonal absolute
daitva	duty or responsibility
dan	contribution; donation; gift
dharma	duty; moral order; religion
dhoti	waistcloth worn by men
dhuni	sacred fire pit associated with wandering renouncers
diksha	initiation by a guru that usually involves obtaining a mantra
divyalok	region of celestial beings
dotara	a lute-like instrument
gaumutra	cow's urine
ghar	house or home of layperson
ghuspeti	infiltrator; spy
guru	spiritual teacher in Hinduism
havan	sacrificial fire
Hindutva	Hindu nationalism
insani batein	human affairs
jati byabasa	caste occupation
jina	literally conqueror; omniscient spiritual teacher of Jainism; synonym with Tirthankara
jiv / a	soul; that which has sentience in Hinduism and Jainism
jnan / a	knowledge; wisdom
jnanamarg	path of knowledge
kamkarnewali	literally she who works
karma	action or good works; fruits of action in Hinduism
karmayoga	the yoga of good works
karmic	pertaining to actions or good works in this or other lives
kashaya	passions
kavatchan	tunic worn by *mumukshus, upasikas*, and *samanis* in Jainism
khepa	crazy; mad with longing for Divine; Bauls may append this adjective to their name
khilaphat	Caliphate; used by Bauls and some Sufis to refer to institution of renunciation
khukuri	traditional Nepali curved knife with an iron blade
kriya yoga	particular yogic technique; used synonymously with *karmayoga* by some renouncers to mean "the yoga of action"
kuti	hermitage; small room, especially in a temple

Lalan panthis	followers of the famous Baul practitioner and artist Lalan Shah who lived during the eighteenth century; his songs, sung by Bauls in West Bengal and Bangladesh, have esoteric meanings that are passed on from guru to disciple after initiation (*diksha*)
linga	phallic representation of Shiva; a sign of Shiva's divine presence and spiritual potential
madhukari	gathering honey by a bee; refers to mendicant practice of taking a little from many households and never a lot from one
mahatma	great soul; term of address usually for (male) saints
mahavratas	great vows of a renouncer
majzub	person who has renounced worldly concerns as a result of losing her or his senses; having ecstatic devotion or divine possession
mala	garland or necklace; rosary used in recitation of a mantra
maya	illusion; attachments and affections one feels for others
mazar	Muslim shrine dedicated to a saint
mithini	ritual sister; fictive kinship bond established between two women in Nepal
mithyadrishti	state of ignorance; delusion
moksha	enlightenment; state of spiritual emancipation
muhpatti	mouth covering worn by Jain renouncers
mumukshu	one who desires emancipation; term used to describe a first year student at school for monastic training
muni	male renouncer
murti	icon in which a deity resides
murti puja	devotional worship of icons
nigodas	the simplest form of life in Jainism
nirguna	without attributes or qualities
nirjara	the elimination of karmas in Jainism
pagal	crazy; see *khepa*
pandit	Brahmin priest
paramlok	heavenly region
path	recitation
pran / a	life force; breath
prasad	gift from a deity or holy place; blessed food
pratikraman	a ritualized confessional prayer in Jainism
pratilekhna	the careful checking of clothes, books, and so on for insects
pravachan	religious sermon
puja	devotional ritual in Hinduism; veneration
pujari	ritual officiate

rajya	rule; kingdom
roti	unleavened flatbread made from wheat
rup	beauty; form; color; menstrual blood used in Baul initiation rites
sadhana	religious or ritual discipline; spiritual practice
sadhu (*f. sadhvi*)	Hindu renouncer; ascetic
saguna	with attributes or qualities
sallekhana	ritual fast to death in Jainism
samaj	society; community
samani	semi-renouncer category within Jain Terapanthi sect; woman who is initiated but permitted some practices forbidden to "full" renouncers, such as traveling by means other than by foot
samayik	meditation on true self; state of spiritual equanimity
samsar / a	cycle of birth, death, and rebirth; worldly existence
samskar / a	mental impression forming on the mind; habit; also life cycle rite for Hindu householders
sannyas / a	renunciation of householder life for other wordly pursuits
sannyasi (f. sannyasini)	Hindu renouncer
seva	service
shakti	female creative principle; strength
shiksha	learning; education
silpi	artist; performer; craftsperson
sindhur	red powder worn by women to symbolize marital status
sthan	place; location
sudarshan chakra	sacred discus associated with Vishnu
Sufi	Muslim mystic
Sufism	mystical form of Islam emphasizing inner meanings of Muslim texts, mystical states to achieve closeness to God, and the importance of a guide (*pir* or *mursid*) to gain knowledge; popular form of Islam in South Asia
swami	lord or master; term of address for male sadhus
tabla	drum
tapas / ya	ascetic discipline, austerities
tirthankara	omniscient spiritual teacher of Jainism; synonym with *jina*
tulsi mala	garland made of carved wood from a sacred basil plant worn by Vaishnavas

tyagi(f. tyagini)	a renouncer; one who has relinquished wordly concerns
udasi	free from worldly constraints; carefree
upasika	worshipper; term used to describe girls who are preparing to take initiation into renouncer order
vairag / ya	detachment; desirelessness
vairagi (*f. vairagini*)	a renouncer who is in a state of detachment
Vaishnava	worshipper of god Vishnu or of one of his incarnations
vanaprastha	forest dwelling; one of the four stages of life in classical Hinduism
vandana	devotional salutation
vipassana	a meditative technique
vritti	inner disposition
vyavahar	wordly behavior; business
yogi *(f. yogini)*	practitioner of yoga; spiritual practitioner in the Hindu or Buddhist tradition
yoni	female organ of generation; seat, place of rest, home, womb; usually represented at base of Shiva linga icon
zildawar	revenue official in Jammu and Kashmir in the colonial era

Tibetan Language

a che	elder sister
a ni	aunt
a phyi	grandmother
be le pang byes	expiatory rite in which ransom figures (*be le*) are tossed
Bon	Tibetan religious tradition that predates Buddhism and is still practiced in areas of Tibet and the Himalayas. Now recognized by the Dalai Lama as a fifth sect of Buddhism
Bonpo	practitioner of the Bon tradition (Tib. Bon po)
brgya 'bzhi	expiatory rite involving the ritual tossing of effigies
'bum skor	village ritual involving circumambulation of the fields
chomo	female religious practitioner of the Tibetan Buddhist and Bon traditions who lives independently in a village, without monastic support (Tib. *jo mo*)

dbu mdzad	chant-master; the monk or nun responsible for leading the monastic assembly in chanting
dge skos	a monastic office of disciplinarian
dge tsha	village ritual to honor the ancestors
dgu mig bzlod	expiatory rite involving the tossing of nine effigies
'dri chang	begging for beer; event to raise funds for rituals
drogmo	ritual sister; fictive kinship bond established between two women (Tib. *grogs mo*)
dzo	yak/cow crossbreed animal used for transport and farm work (Tib. *mdzo*)
geshe	highest qualification in Tibetan monastic system, attained after a minimum of nine years of study and thought comparable to a PhD (Tib. *dge bshes*)
gnyer pa	steward who organizes funding, food, and finances for a ritual or institution
gompa	a temple or monastery in the Buddhist or Bon tradition (Tib. *dgon pa*)
Kagyu	Tibetan Buddhist sect known for its emphasis on meditation (Tib. *bKa' rgyud*)
klu	underground spirits that rule over fertility and prosperity of homes, villages, individuals
mchod gnas	monks serving three-year terms as village ritual officiants
mkhan po	abbot; head of assembly
mnga' 'og	ritual, financial, and legal domain of a monastery within which it has jurisdiction
no mo	younger sister
Nyingma	literally "the old one"; earliest sect of Tibetan Buddhism that shares many practices with Bon and is known for noncelibate householder lamas (Tib. *rnying ma*)
o ma'i rin	literally milk price, paid to a bride's mother
pe rag	jeweled headdress passed on from mother to (usually eldest) daughter at time of her wedding (Tib. *be rag*)
rdo rje slop dpon	a monastic office, head ritual officiant of monastics
rgyal mdos	expiatory rite involving the tossing of thread crosses
ri la bzhugs byes	literally dwelling on the cliff; refers to a nuns' admission into assembly and more permanent residence at the clifftop nunnery
Sakya	Tibetan Buddhist sect widespread in the Mustang region of Nepal (Tib. *sa skya*)
sgrol chu	first watering; laborious process by which fields of seedlings are slowly flooded to prevent soil erosion

skal ba	the share that each nun or monk earns when public donations are divided among members of an assembly
Tamang	major ethnic population living in Nepal's middle hills; they speak a Tibeto-Burman language, practice their own form of Buddhism, and maintain strong links to ethnic Tibetan populations elsewhere in the Himalayas.
tsa mo	granddaughter; niece (Tib. *tsha mo*)
tulku	reincarnate lama in the Tibetan religious tradition (Tib. *sprul sku*)

NOTES ON CONTRIBUTORS

Ann Grodzins Gold is Professor of Religion and Anthropology at Syracuse University. Her research in North India has included studies of pilgrimage, world renunciation, women's expressive traditions, and memories of environmental change. Her publications include four books: *Fruitful Journeys: The Ways of Rajasthani Pilgrims; A Carnival of Parting: The Tales of King Bharthari and King Gopi Chand; Listen to the Heron's Words: Reimagining Gender and Kinship in North India* (coauthored with Gloria Raheja); and most recently *In the Time of Trees and Sorrows: Nature, Power and Memory in Rajasthan* (coauthored with Bhoju Ram Gujar), which in 2004 was awarded the Ananda Kentish Coomaraswamy Book Prize from the Association for Asian Studies.

Kim Gutschow is a Visiting Assistant Professor in the Department of Religion at Williams College. She is the author of *Being A Buddhist Nun: The Struggle for Enlightenment in the Himalayas* (Harvard University Press, 2004), which won the 2005 Sharon Stephens Prize for best ethnography awarded biannually by the American Ethnological Association. She has also written numerous articles on gendering Buddhist monasticism, fasting rituals, Tibetan medicine, pilgrimage, and the politics of water. She has worked in the Himalayan regions of Jammu and Kashmir for 15 years, researching the status of women and nuns, the female body, birthing, postpartum practices, and the relationship between Buddhist rituals, social power, and irrigation practices.

Kristin Hanssen received her doctorate from the University of Oslo, Norway, for a thesis titled *Seeds in Motion: Thoughts, Feelings and the Significance of Social Ties as Invoked by a Family of Vaishnava Mendicant Renouncers in Bengal.* She has worked extensively with Vaishnava Bauls in rural West Bengal, focusing on everyday activities involving singing songs for alms viewed in light of subjective experiences of menstrual blood as a source of physical and spiritual well-being. Her current interests include material culture, popular and political Islam, and issues of gender and modernity.

Sondra L. Hausner is an independent scholar researching gender, migration, and religion as a social system in South Asia. Her manuscript *Wandering in Place: The Social World of Hindu Renunciation* won the 2004 Joseph W. Elder Prize in the Indian Social Sciences from the American Institute of Indian Studies and will be published by Indiana University Press.

Meena Khandelwal is Assistant Professor of Anthropology and Women's Studies at the University of Iowa. She has published several articles and an ethnography entitled *Women in Ochre Robes: Gendering Hindu Renunciation* (State University of New York Press, 2004). She is pursuing research interests in the areas of feminist anthropology, Indian diaspora, and transnational dimensions of Hindu renunciation.

Lisa I. Knight is Assistant Professor in the Departments of Religion and Asian Studies at Furman University. Her research, funded by Fulbright and American Institute of Bangladesh Studies, focuses on how Baul women in West Bengal and Bangladesh negotiate between their identity as Bauls who challenge normative Hindu and Muslim society and their reputation as good Bengali women in the rural communities in which they live. She is currently working on a manuscript, entitled *Negotiated Identities, Engendered Lives: Baul Women in West Bengal and Bangladesh.*

Kalyani Devaki Menon is an Assistant Professor in the Department of Religious Studies at DePaul University. Her research has focused on women in the Hindu nationalist movement and ties into her broader interests in religious politics, gender, and modernity in India. She is currently working on a manuscript entitled *Dissonant Subjects: Women in the Hindu Nationalist Movement in India*, and has published articles based on this research including one that appeared in the *Journal of Asian Studies* in February 2005.

Sara Shneiderman is a PhD candidate in Anthropology at Cornell University. She has worked in Nepal since 1994, and has published on a range of topics including women's religious identity, the ritual production of ethnic identity, and the effects of the Maoist movement on rural political consciousness. Her current research ventures across the Nepali border to India (Darjeeling, West Bengal) and China (Tibetan Autonomous Region) to focus comparatively on the construction of cultural, religious, and political identity in multiple national contexts for the Thangmi, a marginalized ethnic community with populations in all three countries.

Anne Vallely is Assistant Professor in the Department of Classics and Religious Studies at the University of Ottawa with primary research interests in the Anthropology of Religion. Her work focuses on the Jain community, both within India and outside of it, and is particularly concerned with gender, women's religious lives, asceticism, diaspora and transnational studies, ethics, and the symbolic construction of human-nonhuman boundaries. She is the author of several articles and essays on Jain teachings and practices, and a book *Guardians of the Transcendent: An Ethnography of a Jain Ascetic Community* (University of Toronto Press, 2002).

Index

Note: Page numbers in bold refer to illustrations.